PART
THE
COSMIC
VEIL

T0091899

PARTING
THE COSMIC VEIL

Kenneth R. Lang

 Springer

Kenneth R. Lang
Department of Physics and Astronomy
Robinson Hall
Tufts University
Medford, MA, 02155
USA
ken.lang@tufts.edu

Cover illustration: The Orion Nebula, or M17, is a hotbed of newly born stars residing 5,500 light-years away in the constellation Sagittarius. The wavelike patterns of cold hydrogen gas clouds have been sculpted and illuminated by intense ultraviolet radiation from young massive stars, which lie outside the picture to the upper left. The warmed surfaces glow orange and red in this photograph. The green represents an even hotter gas that masks background structures. Various gases represented with color are: sulfur, represented in red; hydrogen, green; and oxygen blue. (Image from the *Hubble Space Telescope*, abbreviated *HST*, courtesy of NASA.)

Title page illustration: A spinning neutron star, or pulsar, at the center of the Crab Nebula is accelerating particles up to the speed of light, and flinging them out into interstellar space. This X-ray image, taken from the *Chandra X-ray Observatory,* shows tilted rings or waves of high-energy particles that appear to have been flung outward from the central pulsar, as well as high-energy jets of particles in a direction perpendicular to the rings. The inner ring is about one light-year across. (Courtesy of NASA/CXO/SAO.)

ISBN-10: 1-4939-0093-5   ISBN: 0-387-33366-5 (eBook)
ISBN-13: 978-1-4939-0093-0

Printed on acid-free paper.

© 2006 Springer Science+Business Media, LLC
Softcover re-print of the Hardcover 1st edition 2006

All rights reserved. This work may not be translated or copied in whole or in part without the written permission of the publisher (Springer Science+Business Media, LLC, 233 Spring Street, New York, NY 10013, USA), except for brief excerpts in connection with reviews or scholarly analysis. Use in connection with any form of information storage and retrieval, electronic adaptation, computer software, or by similar or dissimilar methodology now known or hereafter developed is forbidden. The use in this publication of trade names, trademarks, service marks, and similar terms, even if they are not identified as such, is not to be taken as an expression of opinion as to whether or not they are subject to proprietary rights.

Printed in Singapore   (Apex/KYO)

9 8 7 6 5 4 3 2 1

springer.com

Dedicated to my children,
David, Julia, and Marina,
and
my wife Marcella

Preface

Parting the Cosmic Veil describes our gradual awareness of a much vaster and concealed Universe, more exciting than anyone imagined. It is a story of expanding horizons and the discovery of invisible worlds, made possible with new technology and novel telescopes that have broadened our range of perception and sharpened our vision. They have begun to part the Cosmic Veil, providing a partial glimpse of the Universe in its entirety.

Our account presents those discoveries that have significantly transformed our understanding of the Cosmos. They include the origin of the elements; sending the first humans to the Moon and inquisitive spacecraft throughout the Solar System; observations of planets around other stars; discoveries of the enormous extent and expansion of the Universe; observations of the invisible radio and X-ray Universe, with the discoveries of pulsars, black holes and quasars; the realization that cosmic violence is everywhere, from exploding stars to gamma-ray bursts and the Big Bang itself; and observations of dark invisible matter that fills the spaces between planets or stars and envelops galaxies.

Scientists have one method of perceiving the world, and artists, musicians, poets and writers have other unique points of view. They have all shown us that our awareness is limited by what we are able to see, and, to some extent, by what we expect to see. Some interpret the Universe with line, form, music or words, often illuminating the emotional or mystical perspectives. And just as astronomers use powerful telescopes to transport us into distant realms, beyond our hectic, everyday lives, modern artists also remove us from our immediate surroundings, drawing us into the Cosmos beyond. We therefore broaden our account by including the perceptions of artists, poets and writers, each example chosen for the insight it offers, thereby increasing the appeal and scope of our narrative, but at a modest level that enhances the scientific content of the book and does not interfere with it.

We present this voyage of discovery within universal themes, which provide the book's foundation and explain the chapter titles. They are brave new worlds; motion, content and form; the explosive Universe; the fullness of space; and origins and destinies. There are always unseen worlds that remain to be discovered, with new content and form. Everything is moving, and nothing is at rest. Pervasive outbursts rule the Cosmos, and emptiness is an illusion. All that we can see is evolving, in a perpetual state of impermanence, creation and reorganization. And even though we are pushing the boundaries of knowledge closer to an understanding of the origins and destinies, of either the Universe or Life, the ultimate answers to these grand questions still lie hidden behind the Cosmic Veil.

Although this book is largely an account of our growing realization of the scope and immensity of the Cosmos, it is also a story about people. The fabric of our narrative is therefore bound together with topics that concern us all, from the interests of everyday life to larger questions of our origin, fate and place. Invisibility, motion, content, form, impermanence, explosive outbursts and emptiness, beginnings and ends – these are vehicles for interpreting both the human condition and the Cosmos. Each chapter of this book therefore begins with the human aspects of a theme, helping us to take the Universe personally, followed by the relevant cosmic discoveries.

It indeed seems as if our perception of the Cosmos has been inextricably linked to an evolving understanding of our local personal world. As our ancestors explored the Earth, discovering new continents and seas, astronomers embarked on a similar voyage into uncharted regions of space. It is no accident that cosmic explosions were first understood at about the same time that humans unleashed the forces of the atomic bomb. Theories of human evolution were developed when astronomers were discovering the evolution of the stars and galaxies. Even our current, hurried pace of living seems to coincide with the realization that the galaxies are running away from us at ever increasing speeds, accelerated by a mysterious dark energy. And perhaps our cosmic science owes more than a little to a modern spiritual hunger, a craving for something beyond our familiar, material world.

Parting the Cosmic Veil is written in a concise, light and friendly style that will be appreciated by all, without being unnecessarily weighted down with incomprehensible specialized material. It is broad in scope, but comfortably accessible, a thorough, serious and readable report filled with interesting and informative ideas. There are no obtuse mathematical equations or complex scientific jargon. Throughout this book, scientific concepts have been translated into a common language with apt, down-to-Earth metaphors and analogies, making them accessible to the general reader and adding enjoyment to the material. The text is humanized by interspersing it with personal anecdotes and recollections, and spiced with lively quotations, impressions and reminisces. Vignettes containing historical, literary and artistic material make this book unusual and interesting, augmenting our description of the Cosmos without distracting from the scientific perspective. All of these diverse elements truly set this book apart for the general educated audience, as well as make the material less intimidating.

Numerous images from telescopes on the ground and in space, as well as artists' paintings, are introduced at just the right place within our narrative, providing the mortar that cements our newfound knowledge together. They can crystallize a new concept with a visual excitement that adds a new dimension to our understanding. Many important scientific insights are also compacted within line drawings, with clear labels and thorough captions, which give the reader a concise, forceful message.

Set aside *focus boxes* enhance and amplify the discussion with interesting details, fundamental scientific concepts, and important related topics. They will be read by the especially curious person or serious student, but do not interfere with the general flow of the text and can be bypassed by the general educated reader who wants to follow the main ideas. Equations are kept to a bare minimum, and when employed are placed within the set-aside *focus* elements.

Within each chapter, our story is told in an unfolding chronological narrative, tracing out the growing awareness of a given theme and providing a sense of destination and flow in our gradual parting of the Cosmic Veil. This provides historical authority, and resurrects names and events that have fallen out of time. Our approach emphasizes the ongoing process of astronomy, in which scientists extrapolate from past discoveries in their relentless search for new clues to the mystery of the unknown. And our text does not avoid mention of false assumptions, corrections to missteps in the search, such as the Earth's displacement from its position at the center of the Universe and the discovery of the motions of the fixed stars and fleeing galaxies.

A brief discussion of cosmic vision is presented in the opening chapter, providing an opportunity to introduce our ongoing exploration of the unknown, as well as the artistic perspectives that are lightly peppered throughout the remaining text. This first chapter continues with a discussion of new technologies and novel instruments that have opened new vistas on the Cosmos, many of them a byproduct of military endeavors.

The interdependence of astronomy, technology and the military has now continued for nearly four centuries. It was strengthened during World War II (1939–1945), when the development of the atomic bomb became closely linked to our discoveries of how stars shine and the realization that most of the chemical elements were forged inside stars, during their long evolution and explosive death. And the interdependence continued during the Cold War between the Soviet Union and the United States, when the Americans won the race to put a man on the Moon and space science was begun.

Rockets initially designed to hurl bombs across continents and oceans are now routinely used to loft astronomical telescopes into space, including the *Hubble Space Telescope* that is essentially a spy telescope turned up at the heavens instead of down at the ground. Other aspects of modern astronomy have also benefited enormously from military technology. Detectors of infrared heat radiation are an example, as is the Charge-Coupled Device, or CCD.

Powerful computers and the Internet, originally designed by or for the military, have powerful, far-reaching implications for our celestial science and the rest of the public sector. They have resulted in a new way of doing astronomy, in which multi-million-dollar robotic telescopes, on the ground or in space, are using sensitive electronic detectors and giant computers to create digitized images of the sky.

Our voyage of discovery continues in the second chapter with our familiar *terrra firma* and the discovery that our home planet Earth is a glistening blue and turquoise ball suspended all alone in the chill of outer space. It continues with an account of spacecraft that have carried men to the Moon, and transported cameras and small telescopes throughout the Solar System, obtaining close-up views that have changed the moons and planets from moving points of light to fascinating real worlds that are stranger and more diverse than anyone supposed.

This second chapter concludes with an account of Jupiter-sized and Neptune-sized planets found orbiting nearby stars. This fantastic discovery involved everything that modern technology has to offer, including digital electronic detectors, sophisticated computer software and hardware, optical fibers and improved high-resolution spectrometers to disperse starlight into its separate colors.

Only a century ago, the known Cosmos extended no further than the starry night sky, encompassing all of existence. The Earth was at the bottom of the Universe, and the stars defined its top. Then astronomers built larger optical telescopes to collect more light and see further into space, beyond the stars. In fact, most of the stars in the Milky Way are invisible without the use of a telescope, as are the billions of remote galaxies rushing away from us at speeds that can approach the velocity of light, each with billions of stars wheeling around their massive central hub. They are all spinning, darting and flowing through the Universe, while great concentrations of mass pull the galaxies here and there, distorting the uniform expansion. As far as we can see, the galaxies are also lumped together into vast chains and walls, marking the edges of gigantic seemingly empty voids.

These discoveries, of stars and galaxies with their collective shapes and motions, are described in Chapter 3. Like the discovery of planets around nearby stars, the full distribution of galaxies could not be seen without the aid of electronic detectors, optical fibers, supercomputers, and other marvels of new technology.

Astronomers have now learned to see all over again, using radio waves and X-rays that lie beyond the range of visual perception. They reveal a violent Universe, from exploding stars to the Big Bang, which is presented in the fourth chapter. Here we discover that a massive, bankrupt giant star explodes when it has run out of fuel, while also imploding at its center to form a neutron star and radio pulsar. Precise timing of one radio pulsar indicates that a pair of neutron stars are approaching each other, headed on a collision course and moving together at the rate expected by radiating gravity waves, providing the first evidence for these ripples in the fabric of space-time.

Matter rushing into a stellar black hole, from a nearby visible star in tight orbit around it, emits rapid, irregular bursts of X-rays. Further out in the Cosmos, supermassive black holes power the double jets and lobes of radio galaxies, quasars, and most likely the gamma-ray bursts, all of them radiating an astonishing power never imagined before. Radio astronomers have additionally discovered the relic radiation of the Big Bang, which set the expanding Universe in motion, and used its rippling temperature fluctuations to measure the geometry of the Universe and inventory the dominant kinds of mass it contains.

Our journey continues in Chapter 5, with the discovery that the emptiness of space is an illusion. The space between the planets, which was once thought to be a cold, tranquil and empty void, is filled with hot, charged pieces of the Sun, forming a perpetual solar wind. The Sun also curves and deforms the space around it, twisting Mercury's orbital motion along a winding path and deflecting visible light rays and radio waves that pass near it.

Chapter 5 continues with a description of invisible matter in the unseen places between and beyond the stars. Although most of the interstellar medium is too cold to shine in visible light, it is illuminated by radio waves emitted from abundant hydrogen atoms, as well as a host of molecules, from water and ammonia to formaldehyde and ethyl alcohol.

Cosmic motions have been used to infer the presence of unseen matter on a much grander scale. As also described in the fifth chapter, the outermost stars and gas in some galaxies, including our own, are moving so fast that they have to be held together by substantial amounts of dark matter. The gravitational force of dark, invisible material must also rein in the rapid motions of individual galaxies in clusters. Physicists have additionally speculated that substantial amounts of dark matter have been around since the beginning, to help form galaxies in the early Universe, whose expansion tends to pull the newborn galaxies apart; but nobody has ever found the hypothetical particles.

Seemingly empty space may even be endowed with a capacity beyond the imaginations of many people. Some scientists have speculated that all of the mass and energy in the observable Universe could have originated from the nothingness of space, during a rapid growth spurt, or inflation, soon after the Big Bang, or by a host of microscopic, oscillating strings at earlier times closer to the beginning. These invisible loops of energy are supposed to endlessly and randomly twist and vibrate in a cosmic dance, so perfectly choreographed that they explain everything that exists. Such fascinating speculations are also presented in Chapter 5.

Most recently, the urgency of the "missing" dark matter has diminished, with the discovery that dark energy might take its place. After all, mass is equivalent to energy. As discussed in greater detail in Chapter 5, the dark energy is just as elusive; representing an invisible something that fills the nooks and crannies of supposedly empty space. But no one knows where the stuff came from, or why it might be filling up space.

Chapter 5 concludes with an overview that highlights all the significant things we have learned as astronomers begin to part the Cosmic Veil.

In the closing Chapter 6, and Epilogue, the reader is reminded that everything on the Earth and in the Cosmos is changing from one form to another, and that eventually the Sun and all the other stars will fade away into darkness. This brings us to the mystery of creation, of both the Universe and humans.

Up until recently, science was thought to deal with questions that have answers. Art, poetry and religion were supposed to be about unanswerable questions, and it is the lack of answers to these fundamental questions that gives them power. But now there is overlap between these domains in cosmological investigations of the origin and destiny of the Universe.

A vocal minority of modern cosmologists has proposed highly speculative theories, which might never be verified by observation. They imagine, for example, innumerable Universes so completely disconnected from us, and each other, that we will never ever see them, communicate with them, or know with certainty that they exist. Such theories are comparable to ancient myths, but expressed in the language of mathematics that only the cosmologists can understand.

Although modern astronomy has traced the history of the Universe back to the Big Bang, and thereby explained the origin of the elements, what happened before this beginning is completely unknown, as far as verifiable science is concerned, and may not be understood by science alone.

Scientists can also specify the physical ingredients of living things, and speculate on the conditions for life's beginning and survival. And they are actively searching for life outside the Earth. But no one knows how life originated, or how inanimate, non-biological molecules might have coalesced into a living thing. As with the origin of the Universe, it is possible that an understanding of the origin of life may also lie beyond the capabilities of science alone.

Other speculative topics that may lie outside the domain of pure science include both your fate and the destiny of the entire Universe. Our final sixth chapter also discusses these questions, which now remain beautiful mysteries.

I would like to thank Robert Cole, a student at Tufts University, who read the manuscript for clarity, and NASA's Applied Information Systems Research Program that helped fund the writing of *Parting the Cosmic Veil* through NASA Grant NNG05GB00G.

Kenneth R. Lang
Tufts University *and* Anguilla, B.W. I.
January 1st, 2006

Contents

List of Figures

Chapter One

Cosmic Vision, War and Technology

1.1 NO ONE SEES IT ALL

The Cosmic Veil

Our perception is always limited to just a small part of the much larger world that we are immersed in. We each have our own individual reality, molded by the unique way that we view the external world, and to some extent, we all live in a world that others do not see. So nobody is ever completely aware.

Astronomers observe, measure and quantify the known constituents of the Universe, often looking at them in new ways. Their ultimate goal is to reveal the entire Cosmos. As the American astronomer Cecilia Payne-Gaposchkin asserted, it's all "a search for the Unseen."[1] And in a sense, we only see the shadows on the wall, like people in Plato's cave. Or as William Blake, the English poet and visionary, wrote, "man has closed himself up, til he sees all things thro' chinks of his cavern."[2]

The unseen Cosmos is something like the Japanese rock garden, Ryoan-ji, in Kyoto. It consists of 15 rocks set on raked sand, but only 14 are visible at a time. One rock is always "hidden," and which one it is depends on the viewer's perspective. So the garden has different appearances that change according to the angle of view, or the way of looking. There is always something that remains unseen, something to know more about, but it is either hidden, or viewed dimly from restricted angles or at a distance.

Even our most eminent scientists knew that despite all of their marvelous discoveries they had opened just a small window on a much vaster, concealed Universe. The great English physicist Isaac Newton compared himself to a boy playing on the seashore and diverting himself in "now and then finding a smoother pebble or a prettier shell than ordinary, whilst the great ocean of truth lay all undiscovered before me."[3] And the German physicist Albert Einstein remarked, "all our science, measured against reality, is primitive and childlike."[4]

Einstein knew that there are underlying patterns in the Universe that exist independent of humans, and that we stand before them, awaiting discovery and understanding of a great hidden mystery. To him, anyone who cannot experience that mystery can no longer wonder or feel amazement. Such a person "is as good as dead, a snuffed-out candle."[5]

It is as if the complete essence, the sum total of all things, lies shrouded behind some Cosmic Veil, so the Universe in its entirety, the complete picture show, always escapes our perception. And every method of describing it, including the scientific one, is missing something; all methods of perceiving the world are incomplete.

Ways of Seeing

Everything we see is molded, shaped and constrained by our education, background, and past experience. They determine our individual perception, with which we analyze, interpret and view the world. And it isn't all heredity. Genes count, but differently in different environments, and there is no gene for the human spirit.

Some people are more alert to their surroundings than others, seeing beyond the immediate world around them. Children are a wonderful example. They can have an amazing clarity of vision, with a clean and innocent face that is filled with intelligence. After all, a child looks and recognizes before it can speak, establishing its place in the world.

Young children can be entirely open and receptive, vividly aware of their external surroundings. They can perceive all kinds of fabulous events in a seemingly dull and lifeless landscape. A child might look under the surface of a stream, for example, or notice the sky reflected in a rain puddle. The British poet William Wordsworth remembered such an early childhood experience with:

> There was a time when meadow, grove and stream,
> The earth and every common sight,
> To me did seem
> Appareled in celestial light,
> The glory and the freshness of a dream.[6]

You can detect compassion, intelligence, wonder and all manner of things hidden within a person's eyes. They might be the bright eyes of a child seeing a butterfly for the first time, or the clear, longing eyes of someone in love. Even a person's lively spirit can be mirrored in their sparkling eyes.

And we all have another eye – the mind's eye. We can use it to mentally cast ourselves out of our body, watching it move about in its daily pattern of life. It takes just another leap of the imagination to use your mind's eye to look down on the spherical, rotating Earth, a suspended, spinning ball, or to imagine the Sun hurtling through space and whirling about the distant hub of our Galaxy.

Artists and writers see details that others miss, using their work to show us their perceptions of the world. Paintings, poems, or especially apt metaphors all convey a particular view, often with emotion and feeling. Poets distill a wisdom that comes from the depths of their souls. Artists recreate the world so others might see it anew. And musicians convey other states of awareness or enlightenment. Scientists also create a unique vision of the world, developing a detached perception, an ability to see beyond the apparent. Art, writing, thinking, music, religion, science, and travel – they all help us interpret and understand the world in different ways.

Waking Up, Being Aware

Every so often, the dull mask of daily life shatters like ice, and we become aware of something deeper and truer. It is as if we live within a facade that conceals the underlying fabric of reality from us. But suddenly we wake up and become aware. The English poet and critic Matthew Arnold wrote of this buried life, a hidden self, which occasionally rises up from our internal depths, as from an infinitely distant land. Then:

> A bolt is shot back somewhere in our breast,
> And a lost pulse of feeling stirs again.
> The eye sinks inward, and the heart lies plain,
> And what we mean, we say, and what we would, we know.
> A man becomes aware of his life's flow,
> And hears its winding murmur; and he sees
> The meadows where it glides, the Sun, the breeze.[7]

Most of the time we have a narrower way of seeing and feeling, with blurred vision and cloudy mind, viewing the world through the filter of our selfish interests and the blinkers of everyday habit. There are computers, televisions, and the random perturbations of life that might dull our minds and keep us from thinking. They can tarnish the gleam of the world and dull the sparkle in our eyes, as we look out to see the same old thing, with puffed-up faces that have become marked by the harsher aspects of the world.

But we all wake up from time to time, catching a fleeting glimpse of the beauty that surrounds us. The silence is broken, like the first step on the crust of un-trodden snow, and a pure note that rises out of the confusing noise. Or it might instead be silence, filled with restful potential, like a pause in music that temporarily breaks the melody, or the pregnant quiet between lightning and a thunderclap.

Cosmic Perspectives

Painters use sunlight to illuminate their work, in its direct glare or with slanting shadows. That light can be alive, warm, and life-sustaining, akin to fires that dance in the face of darkness. Or a painting might depict the elusive, varying qualities of the Sun's hot glow, the source of light and color. In religious art, the light often comes from above, representing spiritual power and dividing the celestial from the terrestrial, the heavens from Earth.

The Dutch artist Vincent Van Gogh used thick brush strokes of blazing, brilliant pigment, as dense as honey, portraying another light, a powerful, yellow Sun that blazes forth with an almost supernatural radiance. And for him, the colors did not die at night. To Van Gogh, it often seemed "that the night is more alive and richly colored than the day."[8] Look up at the night sky, and you will also notice that the stars gleam with blue, green, yellow and red colors.

So Van Gogh, the painter of the Sun by day, portrayed the brilliant stars at night, describing their sparkling colors as emeralds, lapis lazuli, rubies, and sapphires. His first *Starry Night* painting, for example, contrasts the blue-green night sky and the huge, sparkling stars of the Big Dipper with the "rough gold" of the lights below (Fig. 1.1). Vincent described the painting in a letter to his brother, declaring "I have a terrible lucidity at moments, these days when nature is so beautiful, I am not conscious of myself any more, and the picture comes to me as in a dream."[9]

Painters are seeking new patterns in the world, using novel perspectives to look beyond the obvious and beneath the surface into the otherwise invisible. The Dutch

FIG. 1.1   Starry night, Arles The Dutch artist Vincent Van Gogh (1853–1890) created this painting, also known as the *Starry Night Over the Rhône*, in September 1888, describing it in a letter to his brother Theo with "The sky is greenish-blue, the water royal blue, the ground mauve. The town is blue and violet, the gas is yellow and the reflections are russet gold down to greenish-bronze. On the blue-green expanse of the sky, the Great Bear [the Big Dipper] sparkles green and pink, its discrete pallor contrasts with the harsh gold of the gas. Two colorful little figures of lovers [are] in the foreground."[a] The painting, can be compared with Van Gogh's more visionary *Starry Night* composed the following year at Saint-Rémy (see Fig. 3.12). Oil on Canvas, 0.724 × 0.92 m (Private Collection, Courtesy of Musées Nationaux, Paris).

graphic artist, Maurits Cornelis Escher, or M. C. Escher for short, was representative, with his skill at portraying spatial perspectives (Fig. 1.2).

Artists can also cultivate a sense of detachment, as if trying to look at everything from a distance to get a larger, more complete vision. You might say that they just want to retain *perspective,* from the Latin for "to look through." And that perspective has changed as astronomers have transformed our understanding of the Cosmos. Art and astronomy have together taken us beyond the curtained heavens of medieval times, into vast, open cosmic space, a world without end, culminating in our modern realization that man is no longer the center of the Universe or the measure of all things.

A poet's vision can also sweep us into the boundless celestial realms, keeping the world forever young; as in this passage, which describes what the Chilean poet Pablo Neruda felt when he wrote his first line of poetry:

[a] Vincent Van Gogh (1853–1890), *The Complete Letters of Vincent Van Gogh.* New York Graphic Society, Geenwich, Connecticut 1958, Volume 3, page 56.

FIG. 1.2   Three worlds The Dutch graphic artist Maurits Cornelius Escher (1898–1972) was a master at depicting various perspectives. In this print, a forest pond is shown with three distinct components. Fallen autumn leaves suggest the surface of the water, which recedes and blends into the sky at an invisible horizon. The reflection of the trees in the distance provides a second component. In the foreground, a fish seems to stare out at us from a third location, below the water surface. 1955 Lithograph 0.362 × 0.247 m. (© 2003 Cordon Art B.V.–Baarn–Holland. All rights reserved.)

And I, tiny being,
drunk with the great starry
void,
likeness, image of
mystery,
felt myself a pure part of the abyss.
I wheeled with the stars,
my heart broke lose on the wind.[10]

Like the poet, astronomers provide us with a sense of something larger than us, which lies beyond our individual daily concerns. And it's at just the right time, when our close, intimate natural world is becoming a dream of the past. Nobody sits in front of their house anymore, to take in the breezes of the afternoon. And practically no one listens to you, or looks you in the eye.

Although we used to look up, to see the stars wheeling slowly across the black lagoon of a clear night sky, tall buildings now block out the celestial heavens and city lights drown out their beacons. Modern astronomy helps fill this vacuum, with remote telescopes that have carried us out through the darkness and shadows, even beyond the known stellar worlds (Fig. 1.3). They have drilled a hole through the heavens, transporting us into captivating realms that lie far beyond our busy little planet with its hectic pace and material concerns.

Modern artists have also been separating themselves from their immediate environment, placing nature at a distance to obtain a cosmic perspective. The Russian painter Wassily Kandinsky was one of the first to become completely detached from his everyday surroundings, stating that "this art creates alongside the real world a new world, which has nothing to do *externally* with reality. It is subordinate *internally* to cosmic laws."[11]

FIG. 1.3 Endless galaxies A remote cluster of galaxies, designated CL 0939 + 4713 for its coordinates on the sky, as it looked about 10 billion years ago when the light we see was emitted and the Universe was two-thirds of its present age. Many of the galaxies have distorted forms, suggesting that mergers and other interactions have disrupted them. This image was taken from the *Hubble Space Telescope* in January 1994. (Courtesy NASA, STScI, and Alan Dressler, Carnegie Institution.)

The painter omitted all resemblance to the immediate physical world in his work, obtaining depth and perception from cosmic space that could continue beyond the borders of his canvas. In some of his paintings, Kandinsky used multiple centers of interest, such as circles of varying size and color, that tend to cancel each other and draw the viewer's eyes in and out, thus creating a fourth dimension in a single throbbing whole (Fig. 1.4). They were used as a cosmic metaphor, especially when painted against a dark background, suggesting planets or stars in the night sky.

So it's perhaps not surprising that astronomy was a source of Kandinsky's inspiration. In fact, he held a lifelong interest in the celestial science. He had an astronomer

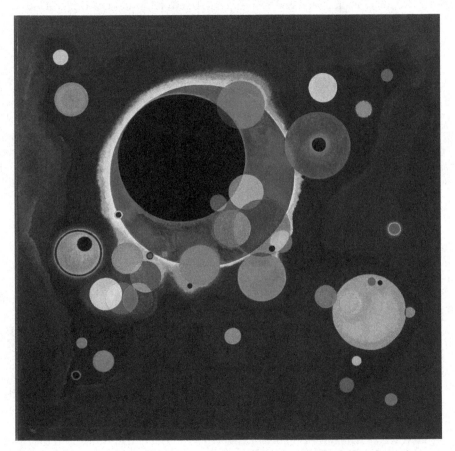

FIG. 1.4 Several circles This cosmic image, painted in 1926 by the Russian artist Wassily Kandinsky (1866–1944), draws us into space, reminding us of unseen worlds that seem to pulsate with energy and vitality. The large blue circle with white edges suggests a cloudy sky, while the deep black circle reminds us of the dark night sky. The transparent intersections of different circles lead the eye into the distance, but without any specification of the depth. It is hard to determine which circles are closer and which are more remote, as the eye focuses on different parts of the painting. Oil on canvas, 1.40 × 1.41 m. (Courtesy of the Solomon R. Guggenheim Museum, New York City, New York.)

come to give regular lectures to his family, when they would take turns looking through a telescope at the Moon, Saturn and the Pleiades star cluster. Kandinsky even once planned to construct a small private observatory in his home in Moscow.[12]

The Spanish painter, Joan Miró, created otherworldly gouaches, known as the "Constellations," which also remove us from our immediate surroundings and draw us into the Cosmos beyond. When Miró created these paintings, between 21 January 1940 and 12 September 1941, the Spanish Civil War (1936 – 1939) had recently ended, the world was falling apart in an enormously destructive World War II (1939 – 1945), and humans were subject to horrible experiences, inner agonies and a loss of identity. So Miró was probably responding to the plight of people at this tragic moment in history. Like many other people at the time, he wanted to flee from the devastating world around him, writing: "I felt a deep desire to escape. I closed myself within myself purposely. The night, music and the stars began to play a major role in suggesting my paintings."[13]

In Miró's work, intersecting circles and lines connect points and mythological signs to the stars, crescent moons and orbiting worlds (Fig. 1.5). The result is a complex web of interrelationships that join human beings, denoted by eyes, to the Cosmos. So he mimicked the stellar constellations, replacing them with his own imaginative vision, and was thus inspired by the celestial heavens.

Einstein agreed with this need to withdraw into a better world, stating "One of the strongest motives that leads men to art and science is to escape from the rawness and monotony of everyday life and take refuge in a world crowded with the images of our own creation."[14]

Open Minds, Serendipity and New Technology

The new discovery, the previously unseen wonder, is everywhere, remaining to be revealed by parting the Cosmic Veil. It might be just a simple thing – a phosphorescent, blue bird or a red dragonfly moving across a lake, or perhaps a veined yellow leaf drifting in the wind. Or it could happen when an astronomer is looking at something else, with an open mind ready to notice the unexpected.

Like creation in other areas, astronomical insight is something like taking a trip to a distant country that you have never seen before. The disorientation turns the ordinary into the unusual and strange. And when alert, you can suddenly notice things that have always been there, sharpening your focus to look at the Cosmos in a new way. Louis Pasteur, the French scientist who developed the germ theory of disease and the technique of pasteurization, expressed it succinctly with "in the field of observation, chance favors only the prepared mind."[15]

But it wasn't scientific brilliance or conceptual foresight that resulted in the major changes in our understanding of the Universe. Most of the significant astronomical discoveries have been accidental, unanticipated and completely serendipitous. And no one predicted that these fascinating discoveries would occur! They are mainly a consequence of new technology and the development of novel telescopes.

These instruments of discovery have extended our vision, enabling us to see the invisible and enlarging our perspective of the unknown. Like an imaginative story, they have taken us to unintended places, each time revealing a new previously hidden aspect of the Universe.

FIG. 1.5 Ciphers and constellations in love with a woman In his
Constellation series, executed between 21 January 1940 and 12 September
1941, the Spanish painter Joan Miró (1893–1983) sought refuge in the
stars, which gave him hope for new life. This painting, created in 1941,
includes a close network of symbols and intersecting and overlapping lines,
set on a nebulous and transparent background. Its parts are interconnected,
as if they were all orbiting around one another or joined like the stars in a
constellation seem to be. Gouache and turpentine paint on paper 0.46 ×
0.38 m (Courtesy of the Art Institute of Chicago, Chicago, Illinois.)

1.2 THE SKY ABOVE AND THE ENEMY BELOW

Telescopes That Detect the Unseen

The instrument-driven voyage of astronomical discovery began with the invention of
telescopes that gather in the colored light seen with our eyes. We sometimes call them
optical telescopes, since the science of optics describes their component lenses or mirrors. They are also known as visible-light telescopes, to distinguish them from radio or
X-ray telescopes that detect invisible radiation.

For centuries after the invention of the optical telescope in 1608, astronomical discovery followed the development of successively larger telescopes, capable of gathering in more light to see fainter objects or resolve bright ones with greater detail. Like the geographic explorers of earlier times, astronomers used these new tools to venture into uncharted territory, finding new worlds that had never been seen before. They discovered new planets, unseen stars and an entire Universe of galaxies that require telescopes to see.

Large optical telescopes pushed the boundaries of the known Universe farther and farther outward, showing that it is much larger and more complex than anyone had anticipated. As a result, we now know that we live on a small rocky planet, orbiting an ordinary star, the Sun, located in the suburban outskirts of an undistinguished sort of spiral galaxy, containing about 100 billion stars, itself just one galaxy among countless others.

These galaxies, which remain unseen with the unaided eye, are grouped into clusters and superclusters that can span 100 million light-years and harbor 100 thousand galaxies (Fig. 1.6). Their shape and form resemble the tangled drip paintings of Jackson Pollock, whose components move in and out of focus as the eye moves back and forth across the canvas, without beginning or end (Fig. 1.7).

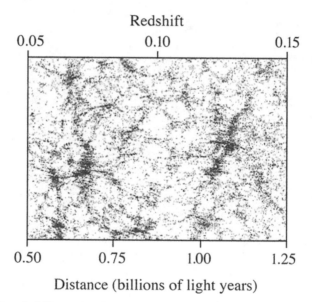

FIG. 1.6   Galaxy bubbles and walls By determining the velocity, or redshift, of tens of thousands of galaxies, astronomers can gauge their approximate distance and determine their three dimensional distribution. Each point on this display, which extends across 750 million light-years, corresponds to one galaxy. It shows that galaxies are located at the junctions and surfaces of immense bubbles, about 100 million light-years across, whose interiors appear to be largely devoid of matter. The galaxies are also concentrated in thin, enormously wide, sheet-like walls. This image corresponds to a strip of the sky that is 2 degrees wide, in declination, and extends across about 15 degrees, or one hour of right ascension. The distance scale is determined from the redshifts, which range between 0.05 and 0.15, corresponding to radial velocities between 15,000 and 45,000 kilometers per second, abbreviated km s[-1]. (Courtesy of the 2dF Galaxy Redshift Survey and the Anglo-Australian Observatory.)

FIG. 1.7   Number 3, 1949: tiger This painting by Jackson Pollock (1912–1956) is comparable to the tangled, filamentary web of distant galaxies (Fig. 1.6). By dripping paint onto an unstretched canvas tacked to the floor, Pollock could work on the painting from four sides, creating a sweeping flow that carries the eye across the canvas, with no beginning or end. (Courtesy of the Hirshhorn Museum and Sculpture Garden, Smithsonian Institution.)

For thousands of years, our telescopic perception of visible light has governed our understanding of the Cosmos, but now we know that there is much more to the Universe. In addition to the rainbow of colors detected with our eyes, there is invisible radiation as well. Every cosmic object emits both forms of radiation, the visible and invisible, but with varying intensity. So some of them can only be detected by observing their invisible rays, and others by their visible light alone. To obtain a complete understanding of the Universe, astronomers now use telescopes that detect all of these emissions, much as doctors examine you with more than their eyes, using stethoscopes, ultrasonic scanners, and X-ray images.

When tuning into the Universe with invisible radio waves and X-rays, astronomers have detected cosmic explosions on a scale hitherto undreamed of, in a realm of pervasive cosmic violence and relentless rampage. This tempestuous, invisible Universe remained unknown until new instruments of discovery were developed, often as byproducts of military technology.

Scientific Benefits of Military Technology

Nearly four centuries ago, the Italian physicist and astronomer Galileo Galilei used a spyglass, or telescope, of his own making to discover four moons revolving around Jupiter and to reveal a host of previously unseen stars in the Milky Way. He was a colorful man, who enjoyed companionship, including a mistress who bore him three children, and he delighted in wine, which he called "light held together by moisture."

Galileo was also a practical man, concerned with increased funding and perhaps an advance in position at the University of Padua, which belonged to the free Republic of Venice. So he anticipated the military advantages of the telescope in the defense of this un-walled city on the sea. In a letter to the doge, or chief magistrate, of Venice, Galileo wrote that the spyglass could spot an approaching enemy in time to dispatch a fleet to engage them at sea, or permit assessment of the enemy's land defenses from a distance.

When Galileo demonstrated the benefits of the telescope to Venetian senators, they promptly doubled his salary and granted him lifelong tenure at Padua. And the process of rewarding professors with military grants and related advances in position continues to this day. War and astronomy have indeed been joined together in a marriage of convenience ever since Galileo's time.

This interdependence of science, technology and the military peaked during World War II (1939 – 1945) when entire nations were mobilized and practically no one was exempt. Some of the best scientific minds in England and the United States joined the war effort, with long-lasting consequences for scientific investigations, from the Earth to the stars and beyond.

Radar was invented to detect enemy aircraft, but it led to the new sciences of both radar and radio astronomy. Sonar was developed to find hostile submarines, but it also resulted in the discovery of underwater volcanoes and the spreading ocean floor. And long-range military rockets, previously used to bomb distant cities, lofted detectors above the obscuring atmosphere to discover cosmic X-ray sources.

Advanced military intelligence technology has been, and continues to be, adopted for astronomical purposes. The Lockheed/Perkin-Elmer Corporation, for example, might have won the contract to build the *Hubble Space Telescope,* abbreviated *HST,* because of their past experience in constructing similar mirrors for an even more lucrative fleet of military spy satellites. More than 20 of these secret telescopes, of roughly *HST* size, were placed in orbit by the United States Air Force between 1970 and 1990 for the purpose of looking down at the ground instead of up to the heavens. They were incorporated in a space-based reconnaissance system, which allowed the United States to navigate through the Cold War with confidence as it tracked the limited military assets of the Soviet Union.

Just one of these spy telescopes was adapted to look out beyond the Earth – the *HST* launched into orbit on 24 April 1990 with a myopic, out-of-focus mirror. The flawed telescope was repaired in December 1993 when a *Space Shuttle* crew installed corrective lenses, and also fixed its flapping solar-cell arrays. Since then, the *HST* has dazzled astronomers and the public alike with cosmic images of unsurpassed clarity, and accomplished some significant unanticipated science. It has looked into the depths of space and time, glimpsing newly formed galaxies when the Universe was less than half its present age, watched super-massive black holes consuming the material around them, and helped astronomers determine how a mysterious "dark energy" has taken over the expansion of the Universe.

Perhaps for reasons of national security, defense contractors, and the military they serve, sometimes know a lot more than they are willing to divulge, like doctors and nurses who hurry down the corridors of hospitals, averting their eyes from the sick, knowing more than they will say. If the technical experience of the military intelligence community was more widely known, many of the problems associated with the *HST* could have been avoided.

NASA and the United States Department of Defense now share the development of advanced launch systems, and the arrangement also includes advances in computers, data transmission, detectors, and image cleaning and data compression. The large, thin, lightweight mirrors of the *James Webb Space Telescope,* abbreviated *JWST* and named for the NASA administrator who guided the agency during the *Apollo* moon landings, are also being developed in partnership between NASA and the Defense Department. When unfolded and rearranged after launch, the 36 mirrors will have a combined area six times larger than the *HST,* gathering more light and seeing farther. It will look back in space and time to observe the first light of the earliest stars and emerging galaxies. Since the faint light is most intense at infrared wavelengths, owing to the rapid motion of distant objects away from us, the *JWST* will also benefit from experience in constructing infrared detectors for spy satellites.

The military initiated development of infrared technology in the 1970s, to sense the heat generated by a target when compared to its surroundings, seeing through dust and smoke and locating the enemy in the total dark. Night-vision goggles with infrared sensors are an example. Infrared detectors have also been deployed in aircraft and satellites to monitor the heat given off by Earth-based missile launchings, missile-cone reentry in the atmosphere, and large concentrations of troops and vehicles on the ground. It is said that they can even determine how many warm bodies are inside an enemy tent.

We all glow in the dark, emitting infrared radiation, but since you don't have infrared eyes, you can't see the warmth of your fellow human beings. You can nevertheless reach out and feel it.

A new window on the Universe opened when some of this military technology became declassified and available to the public. It was used to pioneer infrared astronomy in the 1970s and 1980s, with the unexpected discovery of dusty, planet-forming disks revolving around nearby young stars. More recently, NASA may have even benefited from launch delays of the modern *Space InfraRed Telescope Facility*, or *SIRTF*, proposed in 1979, launched on 25 August 2003, and renamed the *Spitzer Space Telescope*, abbreviated *SST*, after the American astronomer Lyman Spitzer, Jr. It uses a lightweight 0.85-meter (33.5-inch) beryllium mirror, because NASA knew the United States Air Force had made mirrors of beryllium for spy satellites. The infrared detector arrays aboard *SST* were originally developed for missile tracking and improved during the wait for the spacecraft launch. Astronomers improved the sensitivity of the military arrays, permitting *Spitzer's* infrared eyes to discover entire galaxies hidden in dusty enclaves, which no one had ever seen before.

New technology can have enormous social benefits, as well as military and commercial ones. An example is the Charge-Coupled Device, or CCD for short, whose development was at least partially funded by the military-intelligence community interested in using it for reconnaissance purposes. Invented at the Bell Telephone Laboratories in 1969, the CCD produces electrically charged signals from most of the light falling on it. Computers read the signals and are used to create digital images from them.

The military now uses CCDs in high-flying aircraft and spy satellites, sending back images of enemy battlefields and installations. And since CCDs can hold a charge corresponding to variable shades of light, they are also used as imaging devices for digital scanners, fax machines, photocopiers, bar-code readers, and both digital and video cameras.

The CCDs have revolutionized visible-light astronomy as well, extending the range of large, ground-based optical telescopes, shortening the time needed to detect exceedingly faint objects, and rejuvenating smaller telescopes that now have the light collecting power of the largest telescopes of bygone eras. Telescopes equipped with the charge-coupled devices can now record the spectrum of a galaxy in just a few minutes, in contrast to the hour-long exposures required with photographic plates. The *Hubble Space Telescope* uses CCDs to provide pictures of the most distant galaxies ever studied, and they are also found in modern interplanetary missions, such as the *Galileo* mission to Jupiter and the *Cassini-Huygens* mission to Saturn.

Astronomers are now adapting all kinds of technology originally funded and intended for defense purposes. Adaptive optics, for instance, was developed to help focus laser beams that might blast enemy missiles or satellites to pieces. Although these death-ray applications have not been successful, astronomers plan to use adaptive optics to tune out the blurring effects of the Earth's murky atmosphere, correcting for the distortion caused when light passes though the air by using computers to adjust the shape of thin, deformable mirrors. The technique will result in increased angular resolution with ground-based telescopes, improving the clarity of their images.

Additional astronomical benefits were derived from nuclear physics, which led to our detailed understanding of how star's shine. Similar science resulted in the first atomic bombs, and eventually to the discovery that most elements were forged inside stars.

1.3 STELLAR ENERGY, ATOMIC BOMBS, AND THE ORIGIN OF THE ELEMENTS

How Do the Stars Shine?

When your grandparents were young, we didn't know why the Sun shines. Scientists measured the total amount of sunlight that illuminates and warms our globe, and extrapolated back to the Sun, finding that it is emitting an enormous power. And fossil evidence on Earth indicates that the Sun has been warming our planet and shining in this way for more than 3 billion years. No one knew how it could shine so brightly for so long.

The German physicist Albert Einstein provided one clue to the Sun's mysterious source of energy in 1907, when he showed that mass is fully equivalent to energy, meaning that mass loss could be transformed into radiated energy. Roughly a decade later, the English chemist Francis W. Aston provided another clue, by demonstrating that the helium atom is slightly less massive, by a mere 0.7 percent, than the sum of the masses of the four hydrogen atoms that enter into it. In other words, what you get out in synthesizing helium is less than what you put into it.

At the time, astronomers were beginning to think that the stars might be the crucibles in which lighter elements are compounded into heavier ones. The English astronomer Arthur S. Eddington extrapolated Aston's discovery into the Cosmos, suggesting that hydrogen might be transformed into helium within stellar interiors, with the resultant mass difference released as energy to power the stars. His 1920 report included the prescient announcement that "If, indeed, the sub-atomic energy in the stars is being freely used to maintain their great furnaces, it seems to bring a little nearer to fulfillment our dream of controlling this latent power for the well being of the human race – or for its suicide."[16]

The New Zealand-born, British physicist Ernest Rutherford was then using helium ions to bombard atoms, concluding in 1920 that atoms are largely empty space, with most of their mass concentrated in a central nucleus, and that the massive nuclei of all atoms are composed of hydrogen nuclei, which he named protons.

The problem with Eddington's idea of using sub-atomic, or nuclear, energy to make stars shine was that all the protons have the same positive electrical charge and therefore resist coming together. Even at the enormous interior stellar temperatures, the mean velocities of the protons are far smaller than those needed to penetrate this electrical barrier. But Eddington was certain that nuclear energy fueled the stars, remarking, "We do not argue with the critic who urges that the stars are not hot enough for this process; we tell him to go find *a hotter place*."[17]

The paradox was resolved after the Russian physicist George Gamow explained why the nuclei of radioactive substances are releasing energetic particles. He used the uncertain, probabilistic nature of the very small, adopting the quantum theory, to show in 1928 how a sub-atomic particle could be anywhere, everywhere, and nowhere at all. The particle acts like a spread-out thing, existing in a murky state of possibility without a precisely defined position.

This location uncertainty can explain the escape of fast-moving, energetic particles from the nuclei of radioactive atoms like uranium. Physicists like to say that the sub-atomic particles tunnel through the electrical barrier. In this surreal world of sub-atomic probability, one could relentlessly throw a ball against a wall, watching it bounce back countless times, until eventually the ball would go through, or under, the wall.

Within a few years, the Welsh astronomer Robert Atkinson, working at England's Royal Greenwich Observatory, was able to combine Gamow's nuclear penetration probability with the distribution of particle velocities in the hot central portions of stars like the Sun, working out a theory for nuclear energy generation within them. It was immediately clear that the most effective nuclear interactions were those involving light, fast-moving nuclei with low charge, and that only nuclei moving with exceptionally high speed would be able to overcome their mutual electrical repulsion and fuse together. For these reasons, the most likely nuclear reactions within stars involved the collision of protons, the nuclei of the lightest element hydrogen, which would still only occasionally merge and bind together, explaining why stars do not expend all their energy at once and shine for very long times.

By 1931 the great stellar abundance of hydrogen had been established, and Atkinson hit upon the idea that the observed relative abundances of the elements might be explained by the synthesis of heavy nuclei from hydrogen and helium by successive proton captures. But the detailed nuclear reactions inside stars could not be understood until the neutron and positron were discovered, both in 1932, and the relevant nuclear cross sections had been measured at stellar energies.

At the suggestion of the German physicist Carl Friedrich von Weizsacker, the German-born American physicist Hans A. Bethe investigated the fusion of two protons. And then Gamow, who had defected to the United States in 1933, suggested to one of his graduate students, Charles Critchfield, that he calculate the details of the proton-proton reaction. These results were sent to Bethe, who found them correct, and in 1938 the two published a joint paper entitled "The Formation of Deuterons by Proton Combination."

Later that year, Gamow organized a conference in Washington, D.C. to bring astronomers and physicists together to discuss the problem of stellar energy generation. The Swedish astronomer Bengt Strömgren reported that since the Sun was predominantly hydrogen it would have a central temperature of about 15 million kelvin, rather than 40 million as estimated by Eddington under the assumption that the Sun had about the same chemical composition as the Earth. The lower temperature meant that the calculations of Bethe and Critchfield correctly predicted the Sun's luminosity, and Bethe, who attended the conference, was able to explain just how the Sun shines by synthesizing helium nuclei from successive proton collisions. In 1967 Bethe was awarded the Nobel Prize in Physics for his discoveries concerning energy production in stars.

Also in 1938, Weizsäcker independently showed how other nuclear reactions could fuel stars that are more massive than the Sun, using carbon as a catalyst in the synthesis of helium from hydrogen. And in the same year, the German radio chemist Otto Hahn, who had been working with Lise Meitner, showed that when uranium is bombarded with neutrons it could be split into two nearly equal fragments, releasing large amounts of energy. The result was confirmed by Meitner, who was in exile in Copenhagen, and the Danish physicist Neils Bohr described it to Einstein in 1939.

World War II had begun, and scientists feared that Nazi Germany would put Hahn's discovery to use, building an atomic bomb and conquering the world. So Einstein wrote letters to United States President Franklin Roosevelt in 1939 and 1940, encouraging a program be set up to achieve a nuclear chain reaction, and that the United States consider the development of "extremely powerful bombs" that the Germans might also be constructing. And the concern was real, for it was later discovered that Weizsäcker helped the Germans investigate the feasibility of constructing nuclear weapons during World War II (1939 – 1945).

The Shatterer of Worlds

In 1942 the famed American physicist Jules Robert Oppenheimer invited a small group of theoretical physicists, including Bethe, to the University of California at Berkeley to discuss how an atomic bomb might be assembled. And within a year they had all moved to Los Alamos, the secret laboratory in New Mexico where thousands of scientists, technicians and military personnel worked under Oppenheimer's enthusiastic direction to create the first atomic bomb.

Some of the best scientific minds in the country were involved. They included Bethe, head of the Los Alamos theoretical division, Richard Feynman, who worked on numerical calculations of bomb performance, and the Italian immigrant Enrico Fermi, who helped produce the first self-sustaining nuclear chain reaction. Another famous physicist, Philip Morrison, accompanied the first bombs all the way to their final flight, caring for them before they were dropped on Japan.

The world's first nuclear device, a 21-kiloton bomb code named Project Trinity, was detonated on 16 July 1945 at Alamogordo, New Mexico. Witnessing the explosion, Oppenheimer, who was deeply interested in Indian philosophy, apparently recalled the verse in Sanskrit from the *Bhagavad-Gita,* whose English translation reads:

> If there should be in the sky
> A thousand suns risen all at once,
> Such splendor would be
> Of the splendor of that Great Being ...

> I am Time, the mighty cause of world destruction,
> Here come forth to annihilate the worlds.[18]

The last two lines have also been paraphrased as "I am become Death, the shatterer of worlds."

Atomic bombs were dropped on the Japanese cities of Hiroshima and Nagasaki a few weeks later, with devastating consequences. As far as the military was concerned, the bombs were a great success, ending the war and more than justifying the cost of $2 billion in 1945 United States dollars and the employment of 100,000 people.

Oppenheimer became a national hero, appearing on the covers of *Time* and *Life* magazines. But in 1948 he confessed that "in some sort of crude sense, which no vulgarity, no humor, no overstatement can quite extinguish, the physicists have known sin, and this is a knowledge which they cannot lose."[19]

Nevertheless, many physicists subsequently condoned the creation of the more powerful hydrogen bomb as a deterrent to Stalin's threat to the free world. The idea of such a super-bomb apparently originated in 1941, when Enrico Fermi used a stellar analogy to describe how it might be made. He knew that stars derive their energy from the fusion of the lightest element, hydrogen, into helium under extreme heat and compression in their cores, and suggested that a man-made explosion might be used to implode and compact material, creating conditions close enough to those inside a star to start the fusion of deuterium, the heavy form of hydrogen nuclei, and releasing roughly a thousand times more energy than an atomic bomb.

A second nuclear weapons research facility, the Lawrence Livermore Laboratory near Berkeley, California, was created in 1952 to focus on the construction of a hydrogen bomb, and the United States detonated the first fusion explosion, code named "Flora", in the same year, at the Enewetak Atoll, Marshall Islands. It completely vaporized the isle of Elugelab, releasing clouds of radioactive material that drifted across the globe. The Soviet Union achieved such an explosion three years later, and numerous detonations of both atomic and hydrogen bombs have subsequently blasted their enormous destructive capability into our minds (Fig. 1.8).

Then, at the height of the anti-Communist crusade initiated by Senator Joseph R. McCarthy, the United States government tried to remove Oppenheimer's security clearance. At the security hearings in 1954, he was accused of being a Soviet spy, perhaps because of his previous support of radical students, including both his one-time fiancé, his wife, and his brother, who were members of the Communist Party. Bethe defended his former boss, but Edward Teller, then Associate Director at the Lawrence Livermore Laboratory, strongly faulted Oppenheimer's judgment. When asked at the hearing if he considered Oppenheimer a security risk, Teller replied: "I thoroughly disagreed with him in numerous issues and his actions frankly appeared to me confused and complicated. To this extent I feel that I would like to see the vital interests of this country in hands, which I understand better, and therefore trust more."[20]

Oppenheimer lost his security clearance, and Teller lost the respect of many scientists.

Where Did the Chemical Elements Come From?

After the war, some physicists began to apply their wartime knowledge of atomic bombs to an understanding of the origin of the chemical elements, the fundamental constituents of our material Universe. In 1948, for example, Gamow considered the age-old question of just how such material originated. He looked back to the dawn of time, before the stars were born, and proposed that the first elements were formed during the most energetic explosion of them all, the Big Bang that propelled the Universe into expansion.

Gamow knew how elements were being produced during chain reactions in nuclear bombs, and reasoned that similar nuclear reactions built up the chemical elements we see today. And he was right. Only nuclear reactions can produce or transform an element, which is why the alchemists never succeeded.

FIG. 1.8 Christmas island Photograph of the test of a nuclear bomb, code-named *Truckee,* delivered by air 16 kilometers south of Christmas Island (now Kiritimati), Pacific Ocean, on 9 June 1962 during Operation Dominic I. The explosion had a force equivalent to 210 kilotons of TNT, about ten times that of the atomic bomb detonated over Nagaski, Japan, on 9 August 1945. *Truckee* was a prototype test of the W-58 warhead carried on the Polaris A-2 missile and deployed on submarine-launched ballistic missiles. Thermonuclear detonations in the humid Pacific created their own localized weather systems; visible in *Truckee's* cloud are multiple cloud condensation structures known as "bells" and "skirts", as well as the more familiar "cauliflower" structure. The United States Air Force 1352nd Photographic Group, Lookout Mountain Station, took this image from Christmas Island. (From the book *100 Suns* by Michael Light, Alfred A. Knopf, 2003, number 81.)

The relative abundances of elements produced in atomic bombs had been studied during the war in order to determine which nuclear reactions had occurred, and Gamow applied the comparison to cosmic abundances determined by astronomers and other scientists.

Hydrogen, the lightest element, is by far the most abundant substance in most of the stars and interstellar space. Helium is the second most abundant element in the Sun and other stars; followed by heavier elements whose overall amounts generally decrease with increasing atomic weight (Fig. 1.9).

Working with his graduate student Ralph A. Alpher, Gamow proposed that all the elements were produced in a chain of nuclear reactions during the earliest stages of the expanding Universe. They supposed that the material Universe initially consisted of a cosmic "ylem", containing only neutrons at high temperature. Some of these neutrons decayed into protons, and successive captures of neutrons by protons led to the formation of the elements.

FIG. 1.9   Abundance and origin of the elements in the Sun The relative abundance
of the elements in the solar photosphere, plotted as a function of their atomic number, Z, on
a logarithmic scale and normalized to a value a million, million, or 1.0×10^{12}, for hydrogen.
Hydrogen, the lightest and most abundant element in the Sun, was formed about 14 billion
years ago in the immediate aftermath of the Big Bang that led to the expanding Universe.

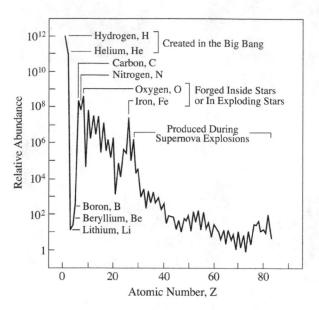

Most of the helium now in
the Sun was also created then.
All the elements heavier than
helium were synthesized in the
interiors of stars that no longer
shine, and then wafted or
blasted into interstellar space
where the Sun subsequently
originated. Carbon, nitrogen,
oxygen and iron were created
over long time intervals during
successive nuclear burning
stages in former stars, and
also during the explosive
death of massive stars. Elements heavier than iron were
produced by neutron capture
reactions during the supernova
explosions of massive stars
that lived and died before the
Sun was born. The atomic
number, Z, is the number of
protons in the nucleus, or
roughly half the atomic weight, since the nucleus of every element except hydrogen contains
as many neutrons as protons, which have a similar mass. The exponential decline of abundance with increasing atomic number and weight can be explained by the rarity of stars that
have evolved to later stages of life. (Data courtesy of Nicolas Grevesse.)

This novel idea was published in 1948, in a paper entitled "The Origin of the
Chemical Elements," with Hans Bethe added as an author, even though he had nothing
to do with the research or the writing, making a pun on the first letters of the Greek
alphabet – alpha, beta and gamma or α, β, and γ – for Alpher, Bethe and Gamow.

Just two years after the α-β-γ paper, the Japanese scientist Chushiro Hayashi
showed that in the first moments of the expansion the temperature was hot enough to
create more exotic particles, such as neutrinos and positrons, the anti-matter particle
of the electron. The mutual interaction of all the sub-atomic particles present in the
first moments of the Big Bang establishes the relative amounts of neutrons and protons,
which in turn determines the amount of helium produced.

With this correction, modern computations by David N. Schramm and others
have conclusively demonstrated that all of the hydrogen and deuterium, and most of
the helium, that are found in the Cosmos today were synthesized in the immediate
aftermath of the Big Bang. So every time you buy a floating balloon, which has been
inflated by helium, you are getting atoms made about 14 billion years ago. And all the
hydrogen atoms found in stars and interstellar space were also created then.

But there are no stable elements of atomic weight 5 or 8, and this seemed to provide an impenetrable obstacle to the synthesis of heavier elements from helium, of atomic weight 4, by the simple scheme of successive collisions with protons, of weight 1. The heavier elements like carbon, oxygen, and iron were instead synthesized inside massive stars that lived and died before the Sun and Earth were born.

The American hydrogen bomb tests in the 1950s indicated that heavy elements are created by rapid neutron bombardment during the bomb explosions, and the American physicist William A. "Willy" Fowler and the English astrophysicist Fred Hoyle showed that similar processes occur in massive exploding stars known as supernovae (Fig. 1.10), explaining the presence of many of the heaviest elements found in the Universe today.

The English husband-and-wife astrophysicists, Geoffrey R. Burbidge and E. Margaret Burbidge teamed up with Fowler and Hoyle to write "Synthesis of the Elements in Stars", published in the *Reviews of Modern Physics* in 1957. Often called the B^{2}FH paper, after the first initials of the last names of the authors, it provided a fundamental framework on which many subsequent studies of nuclear synthesis within stars have been based.

Many heavy elements were formed over long time intervals during successive static burning stages in stars, and the exponential decline in abundance with increasing weight was explained by the rarity of stars that have evolved to later stages of life. Elements heavier than iron, as well as many of the detailed ups and downs of the abundance curve for lighter elements, are attributed to fast nuclear reactions during the explosive, supernova death of massive stars.

So stars are the crucibles where most of the elements were formed, and this discovery was considered so important that Fowler was awarded the 1983 Nobel Prize in Physics – for both his theoretical and experimental studies of stellar nuclear reactions of importance in the formation of the elements in the Universe.

This realization that the elements were forged inside stars connects us to the Cosmos. It seems that we owe our very existence to the cosmic alchemy that took place long ago, in stars that lit up the night sky and were extinguished before the Solar System came into being. The carbon in your molecules and the calcium in your teeth were tossed into space during the explosive death of former stars, as were the oxygen in our water and the iron in the Earth's core, the Eiffel tower and your blood. Joni Mitchell sang about it at the famous August 1969 festival in Woodstock, New York, exclaiming that we are golden stardust.

But all the hydrogen in the Earth's water, and in your body, was made during the earliest moments of the expanding Universe, so we are truly children of both the stars that exploded during past eons and the Big Bang at the beginning of time. These discoveries were a natural byproduct of the development of the atomic bomb, a focused, heavily endowed project of the United States government. The next powerful infusion of governmental funds into astronomy came during the Cold War, resulting in the birth of space science and the landing of men on the Moon.

1.4 THE SPACE AGE

On 4 October 1957, the Soviet Union used an intercontinental ballistic missile to launch the first artificial Earth satellite, *Prosteyshiy Sputnik*, the simplest satellite, trig-

FIG. 1.10   Stellar explosion seeds space This wispy supernova remnant, known
as N49, is located in the Large Magellanic Cloud, or LMC for short, a companion galaxy
to the Milky Way located about 166,000 light-years away and visible from the southern hemisphere of Earth. As the debris from the explosion plows into the surrounding
interstellar medium, the delicate explosive filaments glow in the light of ionized sulfur,
ionized oxygen, and hydrogen. This material will eventually be recycled into building
new generations of stars in the LMC. Our own Sun and planets are constructed from
similar debris of massive stars that exploded in the Milky Way before our Solar System
was formed, some 4.6 billion years ago. An extremely magnetic pulsar, called a magnetar,
is located within the supernova remnant N49, generating repeated outbursts of intense
gamma-ray radiation. This image is about 1.9 minutes of arc, or 90 light-years, wide.
(Courtesy of The Hubble Heritage Team, NASA, STScI and AURA.)

gering a Cold War competition in space. The United States feared that its adversary now had superiority in rocket and space technology that threatened world peace, and decided to fix the imbalance. After all, they reasoned, if you can send a satellite into orbit around the Earth, then you can surely launch a nuclear warhead on a guided missile to anywhere on the planet.

The Americans therefore geared up for action, and in July 1958 established the civilian National Aeronautics and Space Administration, abbreviated NASA, to exercise control over activities in space devoted to peaceful purposes for the benefit of all mankind. Yet, superpower rivalry, national security and the defense of the nation also preoccupied the United States Congress. So the very act that established NASA in 1958 also stated that the Department of Defense would be responsible for aeronautical and space activity "peculiar to or primarily associated with the development of weapons systems, military operations, or the defense of the United States." It was also agreed that there would be a cross-fertilization of civilian and military space interests, and that NASA and the Department of Defense would share information of value or significance to their respective jurisdictions.

From the beginning, NASA's driving force was not based solely on altruistic scientific goals. It was the Cold War rivalry between the two superpowers that led President John F. Kennedy to decide in 1961 that the United States had to surpass the Soviet Union in some dramatic way in space, concluding that they might beat the Soviets in a race to land men on the Moon. Just eight years later, on 20 July 1969, Neil A. Armstrong planted the American flag on the Moon, and the race was won (Fig. 1.11). The rocks returned from the Moon have been used to trace out the satellite's history, and to explain its battered features, but the overriding triumph was political, demonstrating the superiority of American purpose and technology in war or peace.

Former astronauts still send coded messages to men in space, from the NASA facility in Houston, Texas, as they did during the Apollo flights that carried men to the Moon. The control center has no windows, and it is as quiet as the inside of a cathedral, a space-age church. Here, the luminescent screens of computer consoles replace candles and bibles, all facing the green trajectories of spacecraft that snake their way across an altar-like map of the world.

And up above on the fourth floor, it is said, there is an identical control room that is cloaked in secrecy. No visitors go up there. On special days black flags fly near the base of the building, and the control center is strictly off-limits. On those days, the United States Department of Defense sends classified missions into space. One wonders why their flags are colored in black – the color of death, emissaries from the dark side?

The distinctions between the civilian and military objectives in space have now become blurred. Amendments to the 1958 act that created NASA have expanded the definition of space vehicles to include aircraft, missiles, satellites, space transportation systems, including the *Space Shuttle,* and space platforms, such as the *International Space Station.* Many of these vehicles are used for both scientific and defense purposes. In addition, new space technologies, including detectors and telescopes, have been developed and used by both the Department of Defense and NASA, using the same corporations and sometimes with shared responsibilities.

FIG. 1.11   Lunar rover Battery-powered lunar rovers, used in the last three *Apollo* missions, were capable of carrying two astronauts and all their equipment for more than a kilometer across the lunar surface. Because there is no water or weather on the Moon, both the rover and the footprints in the lunar soil may last for millions of years. By that time micrometeorites will have pitted the rover and erased the footprints. (Courtesy of NASA.)

Moreover, the interests of national security can take precedence over NASA's main objective, which is supposed to be science from space. The *Space Shuttle* was, for instance, also designed to put military satellites in orbit, and the Department of Defense has bumped at least one scientific shuttle launch for defense interests, resulting in costly delays in the scientific program affected.

The *Space Shuttle* will most likely continue flying until 2010, as it helps complete the *International Space Station*. Americans will stop using the *Space Station* in 2017, when the total United States contribution will be at least $26 billion. However, no significant scientific results are expected from the *Space Station*. Much more interesting scientific research has been, and will be, accomplished at much lower cost from artificial satellites or space probes launched into space with cheaper, expendable rockets.

So the only justification for the *Space Shuttle* and the *International Space Station* is to retain America's ability to transport astronauts into outer space. But they both resemble public works projects that nourish the aerospace industry. And the politically savvy aerospace contractors, with their extensive lobby pressure on Congress, and the marching armies at NASA centers just will not go away. After all, you don't want to take away anyone's rice bowl.

NASA resources will probably be redirected to new efforts to send humans back to the Moon and later on to Mars. There is serious concern that the funding shift and possible cost overruns will seriously deplete the support of NASA Space Science and hamper scientific research from space.

1.5 SUPERCOMPUTERS, PERSONAL COMPUTERS AND THE INTERNET

Supercomputers provide enormous benefits to the military, astronomy and the society at large. They are now used to transmit trillions of megabytes in real-time – each byte of computer memory stores a single character such as a number or a letter. These supercomputers provide unprecedented knowledge of the exact location of friendly and opposition forces during battle, and tell precisely where to direct precision-guided bombs or missiles, sent from above or beyond enemy fire. In a sense, they've made killing an anonymous, removed and impersonal thing. No single individual is responsible, and almost nobody now stares in the eyes of those who die.

So it is perhaps not surprising that the Defense Advanced Research Projects Agency, abbreviated DARPA, of the United States Department of Defense funded the development of high-performance supercomputers. The United States Department of Energy and National Security Agency also contributed to their development by being very stable customers with fat wallets – the National Security Agency is the military outfit that operates the nation's spy satellites. DARPA has nevertheless been the dominant Research and Development spender for supercomputer development. The acronym for the agency has incidentally changed from ARPA to DARPA to ARPA and now again DARPA, where the D stands for Defense and the name change apparently indicates the dominant governmental interest at the time; but it has been heavily funded throughout by the United States Department of Defense.

DARPA has plans to create a vast computer-based surveillance system, intended to thwart terrorism. The supercomputers might be used to store and sort massive quantities of information that can be used to identify a person's electronic habits; from the phone numbers they dial to their intercepted e-mail messages, inferring their routine habits, organizations and relationships with other people.

But research projects supported by DARPA have not just led to new weapons. They have also fostered commercial technologies such as personal computers, electronic communication, and search engines like Google. There have thus been important, far-reaching implications of information technology initially developed by or for the military. The generals and admirals are fully aware of their ongoing benefits to civilians and private corporations, and have welcomed them. Yet, DARPA initiatives are tending to be less open, with greater focus on "deliverables" that are more directly related to direct military objectives or terrorism threats.

Powerful computers are playing pivotal roles in modern astronomy. They are used to follow targets in the sky, either from the ground or space, and to record, analyze and transmit huge quantities of observed data. Computers are also used to improve the resolution and sensitivity of today's astronomical telescopes by either controlling the shapes of lightweight flexible mirrors or by combining the visible-light radiation or radio waves collected by two or more telescopes to create high-resolution interferometers.

Truly enormous supercomputers are being constructed with government funds. The Columbia Supercomputer at the NASA Ames Research Center in California is one of many examples. It is capable of handling 40 terraflops, which is computer shorthand for 40 trillion floating point operations per second, and that is a lot faster than you can do anything.

They are being used to enhance the scientific capability of the nation, and to also defend it. Live satellite images, for example, might be employed for global environmental protection, by monitoring fires and floods, but also to watch for illegal immigrants along country borders or to describe the terrain of distant battlefields.

Supercomputers are also being applied to reading and understanding immense quantities of astronomical data. Some telescopes in satellites, such as the *Chandra X-ray Observatory*, generate more data than can be analyzed, at least for the time being. Even a decade ago, the *Magellan* spacecraft was using computers and radio waves to penetrate the perpetual clouds covering Venus, taking in 36 million bytes of data every second and relaying them back to Earth for more than four years. As a result, the topography of the entire surface of Venus has been determined, mapping details as small as 120 meters across and producing the most complete global view available for any planet, including Earth.

The Sloan Digital Sky Survey provides another example of the astronomical use of supercomputers and related technology. Optical fibers, charge-coupled detectors and a supercomputer have been attached to a dedicated telescope and used to record the light signals from many hundreds of objects simultaneously. The survey, begun in 1998, is expected to record the brightness and location of more than 100 million stars, and the spectra and distances of greater than a million galaxies. The total quantity of information produced, 30 terabytes (30 trillion bytes and 30 thousand gigabytes), rivals the information content of the United States Library of Congress. It enables astronomers to map the three-dimensional structure of the Universe, determining our place in it and how it has changed with time.

Similar encyclopedic surveys of millions of astronomical objects have been created by dozens of telescopes on the Earth and in space, initiating a new paradigm, a new way of doing astronomy. Multi-million-dollar robotic telescopes, on the ground or in space, are using sensitive electronic detectors and giant computers to create digitized images of the Cosmos from radio waves to visible light and X-rays, and data from literally millions of objects are now being stored in computer warehouses.

Once the astronomical equivalent of a search engine is devised, these resources will be available to anyone with a personal computer and Internet access, regardless of country or institutional affiliation. Virtual computerized observatories will be created, combining surveys with different telescopes operating at various wavelengths. These virtual observatories are the suppositories of a new digital Universe, and scientists are just beginning to use them to find new objects or relationships. They mark the beginning of a new age, when computer screens, available twenty-four hours a day, replace lonely nights peering through telescopes on remote mountaintops.

But to put these efforts in perspective, the spy satellites of the United States National Security Agency now gather as much information in just a few hours as the amount expected from the entire Sloan Digital Sky Survey. Spying on either humans below or the heavens above not only requires supercomputers, but also the right sort of software and other technology to extract significant information from the vast quantities of data that are being obtained. Discovery will naturally follow whenever one of these new technologies is applied to the Cosmos, but there is one aspect of computers that suggests caution. Since they lack the flexibility of the human mind, computers do precisely what you tell them to and nothing more.

Large computers are also now being used in all areas of science as a way of exploring inaccessible processes – phenomena that are too large, too complex, too fast, or too slow to be visualized by traditional experiments. They have helped scientists simulate events that may not be directly observable, and to thereby see the invisible or imagine the inconceivable. An example is Blue Mountain, the supercomputer covering a quarter of an acre at Los Alamos National Laboratory, home of the atomic bomb. This huge calculating machine is used to simulate nuclear explosions without detonating them and endangering Earth. Such computations are also used to describe exploding stars, or supernovae; like atomic bombs they involve the sudden release of great amounts of energy that blows things apart. Similar computations, for both nuclear bombs and exploding stars, are carried out at the Lawrence Livermore Laboratory near Berkeley, California.

High-speed computers and complex theoretical models have traced out the life histories of stars of different mass, from formation through various thermonuclear aging processes. Other computer experiments have been used to study the collisions of black holes, to simulate the clustering of millions of galaxies, and to visualize how the Moon, the planets, the stars and even galaxies might have formed.

On a more practical level, many of us now send daily messages over the Internet, which links computers all over the world and permits individuals to use electronic mail, or e-mail, and the World Wide Web. Nevertheless, it is not widely known that the United States Department of Defense funded the development of the Internet from its inception through the Advanced Research Projects Agency. It operated ARPA net – the precursor of Internet – for the first 10 to 15 years before handing it to other agencies and commercial vendors, and continues to support the development of the Internet.

One of ARPA's original purposes was to design a method of communication from one crucial place, say the White House, to another, like an atomic bomb control center buried under some Colorado mountain, even if entire cities had been destroyed by the enemy. Now the Pentagon is planning its own War Internet in the sky, which will extend and amplify what ARPA began. It will link military aircraft and forces on land or sea to global intelligence, surveillance, reconnaissance and communications satellites in a Global Information Grid.

Nevertheless, the Internet was never used solely for military purposes. Its success has a lot to do with its being free and widely available from the beginning, with the

Department of Defense footing all the bills indirectly through universities and research laboratories.

Tim Berners-Lee invented the World Wide Web, or just the Web for short, in 1989 without the direct support of ARPA. It is essentially a language and protocol for marking up documents so that they can be requested and sent over the Internet. Berners-Lee's vision was that information sharing should be totally voluntary and possible without involving any bureaucracy or outside controls, and he was absolutely right.

The Web is an open, non-proprietary and free method of communicating knowledge that is available to anyone with a computer, cutting across the borders of countries and through elitist institutions and power structures. The distinctions between outsiders and insiders are therefore quickly evaporating as far as information is concerned, and all areas of knowledge are becoming available to any educated, inquisitive person with a computer and Internet access. The Web has also revolutionized the way that people do business, entertain each other, exchange ideas, and socialize with one another, while additionally showing that people love to tell everything about themselves to the whole world.

Our global society has become networked through a worldwide system of computers and satellites. Geosynchronous satellites, that orbit the Earth at the same rate that the planet spins, stay above the same place on Earth to relay and beam down signals used for cellular phones, global positioning systems and internet commerce and data transmission. They can guide automobiles to their destinations, enable aviation and marine navigation, and permit nearly instantaneous money exchange or investment choices. Other satellites move in lower orbits and whip around the planet, scanning air, land and sea for environmental change, terrestrial weather forecasting, or intelligence and military reconnaissance.

When used together with communications satellites, the Internet, the Web and personal computers have brought the whole world together in an immense web of electronic connections, arching around the globe. They provide the veins that channel the flow of cash, political power and criminal connections. They even enable a new kind of personal contact, a sort of online coupling in a digital world that never sleeps. And they also wrap the globe in a network of global surveillance, in which tens of thousands of skilled personnel listen, or read, intercepted messages or scrutinize satellite images.

The Space Age, with its rockets, satellites and computers, has also revolutionized our understanding of the planets. Astronauts bound for the Moon looked back at the Earth, seeing a tiny, fragile oasis suspended all alone in the cold darkness of space. And spacecraft have encountered all the major planets, obtaining close-up views that have changed the moons and planets from moving points of light to fascinating real worlds that are stranger and more diverse than anyone supposed.

Chapter Two
Brave New Worlds

2.1 UNEXPLORED TERRITORY

To our ancestors only a few centuries ago, the ocean, land and sky seemed almost limitless. The Earth consisted of widely separated and isolated continents, each with civilizations that often had little direct contact or knowledge of each other. Then explorers began to look far beyond their local horizons, establishing paths, roads and eventually far-reaching sea routes to previously inaccessible countries. This brought mutual awareness to most of the lands on the once-divided Earth.

The era of European discovery and exploration began in the 15th century, when the Portuguese established trade routes with Africa and India, and Spanish explorers sailed to the New World consisting of the present continents of North and South America – which were new to the Europeans who thought the world consisted only of Europe, Asia and Africa. And for more than a century the Portuguese and Spanish controlled the world's oceans, as their ships, laden with gold, silver and spices, sailed into Lisbon and Cadiz.

At the request of Prince Henry the Navigator, Portuguese ships explored the African coast, searching for gold and silver, taking slaves and hoping to find a westward sea route to India. In 1452, Pope Nicolas V, called "the Great Humanist," issued his papal bull allowing the enslavement of "pagans and infidels."

Four decades later, Christopher Columbus, the Genoese navigator who had long lived in Portugal, changed his allegiance to King Ferdinand and Queen Isabella of Spain, and, with their backing, "sailed the ocean blue" to land at Española, or Hispaniola, and forge a path to the New World. Española is the second largest island in the West Indies, between Cuba and Puerto Rico, now consisting of Haiti and the Dominican Republic.

Within a single generation after 1492, navigators found their way around the entire globe. On 15 August 1519, for example, a fleet of five ships, under the command of the Portuguese nobleman Ferdinand Magellan, set out to find a westward passage to the Spice, or Molucca, islands for the king of Spain. Magellan rounded the tip of South America and reached the Philippine island of Mactan, where he was hacked to death in knee-deep water by natives he had needlessly antagonized. After the death of Magellan, and that of his successor, and the destruction of four of the five ships, the Basque captain Juan Elcano brought the ship Victoria to the Moluccas, sailed around the Cape of Good Hope and returned safely back to Spain on 6 September 1522, completing the first circumnavigation of the world with an emaciated crew and a cargo of cloves that were worth their weight in gold.

In the 16th century, when the renaissance was thriving in Italy with the art of Leonardo da Vinci and Michelangelo Buonarroti, Spain explored and began to colonize New Spain, which consisted of much of South America and the western part of North America within regions now known as Mexico, Peru, California and Texas. Hernando Cortez and Francisco Pizarro respectively conquered and wiped out the Aztec and Inca Empires in the process.

Then for three centuries the British, Dutch, French, Portuguese and Spanish connected the world, opening up distant lands with long-distance sea voyages. Settlements were established, nations conquered, treaties made, slaves captured and sold, and vast trading companies created, all with an eye toward the commercial, the gleam of gold, silk, and silver or the taste of spices, tea, tobacco and rum. The clocks turned, time went on, and eventually it was England that ruled the waves, defeating rival navies and reshaping the world into a far-reaching British Empire. And throughout this time, explorers and sailors looked to the stars to obtain their bearings (Focus 2.1).

Following the American War of Independence (1775–1781), President Thomas Jefferson purchased a huge expanse of western territory – the Louisiana Purchase signed in New Orleans in 1803 – from Napoleon, who was too busy in Europe to be concerned with America. Jefferson then sent an expedition westward, to explore the territory and establish a passage to the Pacific Ocean, with the promise of future access to the silk, spices and gold of Asia. Led by Captain Meriwether Lewis and Lieutenant William Clark, the "Corps of Discovery" traveled 12,800 kilometers in two and a half years, from May 1804 to September 1806, tracing the Missouri river to its origins, crossing the Continental Divide, and reaching the Pacific Ocean.

Within a century, the American West was explored, surveyed and settled by hardy, individualistic pioneers. Five transcontinental railroads had been completed, and westward migration became so rapid that the Bureau of Census declared in 1890 that the American Frontier was at an end. The United States had been created, becoming a great nation of continental proportions that stretched "from sea to shining sea."

By the mid-20th century, British, European and American pioneers had discovered, explored, settled and transformed several New Worlds – New Spain, New Holland, New England, New Orleans and the final New Frontier of the American West. Then the pioneering spirit changed and the New Worlds became old. Eventually there was no more land to conquer or wilderness to tame, and nothing was far away anymore.

Geographers, surveyors and mapmakers specified the exact shapes of the continents and oceans, and the whole Earth became constrained in a quantitative way. Distances part way around the surface were found by the surveying technique of triangulation, and combined to determine the Earth's circumference and a radius of about 6,400 kilometers. You just had to measure one side of a triangle and the two corner angles to get all of its dimensions, and bigger triangles could be constructed from the adjacent smaller one.

FOCUS 2.1
Locations on Planet Earth

For centuries, people have used the stars to find their way across trackless deserts and the featureless ocean. The Big Dipper can, for example, be used to find the bright star Polaris, located above the North Pole, which is in line with the two stars on the cup part of the dipper. And once you've located that star, you can tell which way North is. Just find the part of sky directly over your head, and trace an imaginary line from there to Polaris; the line points North. During the American Civil War, that is how fugitive slaves located their escape route from the south, and that's the meaning of the refrain "follow the drinking gourd."

A grid of great circles, inscribed on an imaginary spherical Earth, is used to define positions at land or sea. A great circle divides the sphere in half, and the name comes from the fact that no greater circles can be drawn on a sphere. As every school child should know, and many of us have forgotten, the two coordinates that establish a position on Earth are known as latitude and longitude.

The latitude great circles are called parallels, since they stay parallel to each other as they girdle the globe. The zero point of latitude is located along the equator, a great circle halfway between the North and South Pole; it is called the equator because it is equally distant between both poles. And all the other latitudes are marked off in a series of shrinking concentric rings, measured northward (positive) or southward (negative) along the circle of longitude from the equator to the surface point.

Ancient navigators determined a ship's latitude by measuring the altitude above the horizon of the Sun at noon. At zero latitude, on the equator, the Sun passes almost directly overhead, while it makes an inclined arc across the sky at other locations.

The meridians of longitude run perpendicular to the equator and any other latitude.

They loop around the Earth from pole to pole and back, in great circles of the same size, converging at the North and South Poles. Any of these circles could serve as zero-degree longitude, but since 1884 this Prime Meridian has been legislated as the half-circle that passes through the Greenwich Observatory in England.

In the late 18th and early 19th century, reliable clocks, known as marine chronometers, were used to determine accurate longitudes at sea, in all kinds of weather and on tossing ships where telescopes faltered and sailors lost sight of land. The clocks kept the Greenwich Mean Time, abbreviated GMT, and could be compared with the local time to determine longitude. Since the Earth takes 24 hours to complete one full turn of 360 degrees, one hour of time difference marks one 24th of a rotation, or fifteen degrees of longitude.

The marine chronometers were eventually replaced in the 20th century by radio transmissions of time signals, such as the Loran-C transmitters employed by the United States Navy and Coast Guard. Atomic clocks, which have a precision of up to a billionth of a second, were then placed within artificial satellites by the United States military, who used them to create the Global Positioning System, or GPS for short. Each orbiting clock transmits specially coded signals that can be used with portable receivers to determine the exact location and time. Although the system was funded and operated by the United States military in the 1990s, and used by them to target bombs and guide missiles, it has since moved into the civilian sector.

Nowadays, such global positioning devices can be used to guide your automobile or to determine your location when exploring distant lands. But the techniques were developed to accurately guide weapons of war, and they are still used for that purpose, with startling pinpoint accuracy.

FOCUS 2.2
Mass of the Earth

The first person to weigh the Earth was the aristocratic, secretive and reclusive English scientist Henry Cavendish, who avoided human contact and made almost no attempt to use a fortune of the order of a million pounds bequeathed to him. He preferred to live in elegant solitude, devoting his entire life to the acquisition of knowledge and transforming his large house into a laboratory where he investigated electricity, gravity, heat, and the composition of the atmosphere and water. He was, for example, the first person to demonstrate the existence of hydrogen as a substance, and the first to combine hydrogen and oxygen to form water.

But Cavendish did not permit all his results to be known. In the last 25 years of his life, he published only five papers, and the most important of these appeared in 1798, when he announced the determination of Newton's gravitational constant, thereby deriving the mass of the Earth. This was accomplished using an ingenious apparatus left to Cavendish by the English clergyman John Michell on his death. Michell was incidentally the first to propose that the gravitational pull of a star might be so strong that no light could escape from it, creating what we now call a black hole.

His measuring device consisted of two small lead spheres attached at the end of a delicate suspended rod, and two larger, fixed spheres. Cavendish placed the equipment in an isolated room, where nothing could disturb it, not even a whisper, and spent a year in exact, painstaking measurements of the twist produced by the bigger spheres on the rod, using a telescope aimed through a peephole in an adjoining room. Newton's gravitational constant was inferred from the twist, and used by Cavendish to determine a mass density of the Earth of about 5.5 times that of water and a mass of about 6 billion trillion tons, or 6×10^{21} tons and 6×10^{24} kilograms for the Earth.

By 1798, the English scientist Henry Cavendish had even measured the mass of the Earth, by determining the pull of gravity between lead spheres, reporting that our planet weighed in at about 6 billion, trillion tons (Focus 2.2). And that's within one percent of modern determinations. So it seemed that there wasn't much more to know in the most basic quantitative way about planet Earth.

As the centuries passed, travelers also began to feel the smallness of the world, and it's getting smaller all the time. Exotic India, for example, was once distant enough to keep the country special. Now we can board an airplane at a whim, flying to India or almost anywhere in the world in less than a day. The entire globe has become interconnected, networked by a vast system of airplanes, artificial satellites, computers, email, Internet, and television. So the world has become our playground, and it has lost many of its exciting, unique qualities as a result. There is no foreign destination that isn't filled with gawking tourists, no new frontier or unexplored territory.

But there were new planets to measure and explore, located outside the Earth and viewed only with an astronomer's telescope. William Herschel, a professional musician and self-taught amateur astronomer, accidentally discovered the first of these brave new worlds, Uranus, in 1781, while sweeping the celestial heavens with a home made reflecting telescope. Its metal mirror was just 0.15-meter (6.2-inches) in diameter. Herschel became famous, almost overnight, as the first man in recorded history to have discovered a new planet. When a previously unknown radioactive element was isolated, in 1789, it was named Uranium, in honor of the discovery. And a few decades later, the young poet John Keats compared Herschel's feat to those of Ulysses, the greatest explorer of Greek mythology, and to the Spanish explorer Hernando Cortez:

> Then I felt like some watcher of the skies
> When a new planet swims into his ken;
> Or like stout Cortez when with eagle eyes
> He stared at the Pacific – and all his men
> Looked at each other with a wild surmise –
> Silent, upon a peak in Darien.[21]

The first asteroid was discovered on 1 January 1801, the first night of the 19th century, and several more soon thereafter. Neptune was found in 1846, at a location predicted from its gravitational perturbations of the motion of Uranus. Tiny Pluto was next discovered, in 1930, but like the asteroids relegated to minor planet status in 1985 when its tiny size and mass were measured using the orbital properties of it oversized moon, Charon. So giant Neptune remains the outermost major planet in the Solar System, serving as a lonely, distant sentinel to outer space.

Numerous small worlds have been found orbiting the Sun just outside Neptune's orbit, tiny balls of ice and rock that are millions of times fainter than can be seen with the unaided eye. All of these brave new worlds – Uranus, a multitude of asteroids, Neptune, Pluto and its companions in the outer fringes of the Solar System – were unknown to the ancients. They are invisible worlds seen only with the aid of a telescope.

Moreover, the distant giant planets remained nothing more than a blurred smudge when viewed with the best telescopes on Earth. They only became known in detail when the inquisitive eyes of robotic spacecraft were used to obtain close-up views, revealing fascinating and unexpected realms that are beyond the range of vision from Earth. This incredible voyage into space began with our Moon.

The silver light of the Moon has beckoned humankind since ancient times, pulling us closer to the heavens and out into the Cosmos. We can almost feel its attractive power, especially when looking at the full Moon near the horizon (Fig. 2.1). The Swiss painter Paul Klee has portrayed the full Moon as a tranquil disk suspended in cosmic space – shining from afar, all alone and free of terrestrial restraints (Fig. 2.2). And the Moon can also seem so close that we can practically reach out and touch it, as in the Mexican artist Rufino Tamayo's portrayal of two women leaping free of the Earth and reaching for the Moon (Fig. 2.3). The Moon is always apart from the Earth and inextricably linked to it, like husband and wife.

2.2 RACE TO THE MOON

About a century ago, on 17 December 1903, the Wright brothers from Ohio – Wilbur and Orville – piloted the first sustained, controlled flight in a powered flying machine. Within half a century, commercial airplanes were carrying paying passengers between large cities, and humans had entered the air. Like birds, we discovered a new dimension, a new freedom through flight in space.

The Soviet Union triggered the Space Age about half a century after the Wright's pioneering flight, launching the first artificial Earth satellite, *Prosteyshiy Sputnik,* the simplest satellite, in 1957, hurling the *Luna 3* spacecraft past the invisible far side of the

FIG. 2.1   An enormous Moon In this awesome picture, a man and child seem enveloped by the full Moon, which looks huge in comparison to the tree in the foreground. When the Moon is overhead, alone in an otherwise empty sky, there are no other objects to gauge its distance; we then think the Moon is smaller than at the horizon. (Courtesy of der Foto-Treff.)

Moon two years later, and sending the first human, cosmonaut Yuri A. Gagarin, into orbit aboard the *Vostok 1* capsule two years after that.

Soviet officials cited their early accomplishments as evidence that Communism is a superior form of social and economic organization. And the United States feared that a missile gap existed between it and its adversary, which seemed to verify the threat that the Soviet Union posed to world peace.

Stimulated by the worldwide excitement generated by the first human fight in space, the brave, visionary young President, John Fitzgerald Kennedy, decided that the United States had to defeat the Soviets at their own game, and deliver an American to the Moon. Thus, on 25 May 1961, just six weeks after the Gagarin flight, Kennedy delivered his now-famous address to a joint session of Congress, including the declaration:

> I believe that this nation should commit itself to achieving the goal, before the decade is out, of landing a man on the Moon and returning him safely to Earth. No single space project in this period will be more exciting, or more impressive to mankind, or more important for the long-range exploration of space; and none will be so difficult or expensive to accomplish.[22]

FIG. 2.2 Full Moon In this 1919 painting by the Swiss artist Paul Klee (1879–1940), a full Moon lies suspended in the Cosmos, its gleaming light illuminating the Earth below. Rectangular objects in the foreground act like a foundation for trees and a triangular form that point up, carrying our eye toward the Moon. It is as if the attractive power of the Moon is pulling the foreground objects and the viewer into space, but that force seems balanced by the Earth's own attraction, perhaps symbolized by the mysterious black circle within the triangle. Oil on paper 0.439 × 0.37 m. (© Joachim Blauel–ARTOTHEK München, Pinakothek der Moderne.)

So the Moon became the new American frontier and the United States space program was galvanized under the newly created National Aeronautics and Space Administration, abbreviated NASA. In less than nine months, John Glenn, Jr. became the first American to orbit the Earth, and the race to the Moon was in full tilt.

FIG. 2.3   Women reaching for the Moon (Mujeres alcanzandola
luna) In this painting, created in 1946 by the Mexican artist Rufino
Tamayo (1899–1991), two women stretch upward in an attempt to touch
the Moon. The angular, elongated forms leap up to nearly ascend to
the Moon, attempting to break free of the Earth's gravity, and perhaps
symbolizing our longing for the eternal. It is as if they were reaching out
to grasp, hold and encompass the Cosmos, also represented by stellar
constellations placed in the space beyond them. Oil on canvas, 0.915
× 0.66 m. (Courtesy of the Cleveland Museum of Art, Donated by the
Hanna Fund, 1947.69 © Estate of Rufino Tamayo.)

Before the United States accomplished manned landings on the Moon, three
types of robot spacecraft were sent to reconnoiter our satellite and answer two main
questions for the proposed lunar landing. The first concerned the danger of encountering rocky terrain, where it would be impossible to land without capsizing. The

second was the prediction, by some astronomers, that a thick layer of dust covers the lunar surface, perhaps as deep as a kilometer, which would make travel impossible. In fact, the astronauts might plunge into the dust and be buried within it, suffocating and vanishing without a trace, like sinking into quicksand on Earth.

To start resolving these uncertainties, three *Ranger* spacecraft crashed into the Moon, transmitting television pictures back to Earth as they rapidly approached the lunar surface. Watching these pictures was a dizzying experience, and the transmission of the final frames was interrupted by the crash itself. These were followed by five *Lunar Orbiters* that mapped most of the Moon's surface to locate potential landing sites, missing only the polar regions. The final stage of preparation involved soft landings by the *Surveyor* spacecraft that tested the detailed physical and chemical properties of the lunar surface and certified the safety of the initial *Apollo* landing sites. While the ground-control crews watched anxiously, the feet of the spidery three-legged *Surveyors* sank only a few centimeters into the lunar soil, showing that there was no thick dust layer and people could, indeed, walk on the Moon without sinking in over their heads.

The *Apollo* spacecraft was designed to carry three men into orbit around the Moon. A small, spartan landing craft, the *Lunar Excursion Module* or *LEM* for short, would ferry two of the crewmen from lunar obit to the Moon's surface and then back to the mother ship, while the third astronaut remained orbiting the Moon in the larger *Command Service Module*.

On 21 December 1968 three *Apollo 8* astronauts became the first humans to break free of the Earth's gravity. Although the crew would only orbit the Moon and not land on it, the astronauts looked back home to see the Earth from afar, and its sheer isolation became plain to every person on the planet.

We then saw our world in a new perspective, as a tiny, vulnerable oasis suspended all alone in the dark void of space. And for the first time we saw our atmosphere for what it is, a thin, fragile membrane. The air that warms and protects us, and permits us to breath, is no thicker, relative to the planet, than the film of tears that clouds the eyes when a lover departs or a child starts to cry.

On 20 July 1969, the spindly-legged, *Lunar Module Eagle* carried two *Apollo 11* astronauts to the shores of the Moon's Sea of Tranquillity. While an estimated 600 million people watched on television, Neil A. Armstrong took the controls to avoid a hazardous crater, and radioed the first words from another world "Houston, Tranquillity Base here. The *Eagle* has landed." It was enough to take your breath away.

With Buzz Aldrin at his heels, Armstrong groped cautiously down the lander's ladder, stood firmly on the fine-grained lunar surface, and planted an American flag on the Moon. An ancient dream had come true – man had set foot on another world and humans were no longer confined to their native planet.

As Armstrong dramatically put it: "That's one small step for a man, one giant leap for mankind," but his words didn't sound very natural. They had probably been produced down in Texas for his use when he set foot on the Moon. The next day, the Italian newspapers put it more succinctly: "Fantastico!"

By the time of the *Apollo 17* mission in 1972, the lunar landings had become so routine that astronaut Eugene A. Cernan muttered "let's get this mother out of here" as he blasted off the Moon in his *Lunar Module Challenger*. That's more like Sir Edmund

Hillary's comment when descending from the summit of Mount Everest on 29 May 1953 – "we knocked the bastard off."

Still, the first human visit to the Moon was a historic occasion. Even now, there is a sense of participation, a feeling that our lives were enriched and made memorable by the landing.

The astronauts' cameras have recorded an eerie wasteland below a blackened sky, battered and scarred with craters of all sizes and covered with dust (Fig. 2.4). It clung to the astronauts' clothing and equipment and showed the sharp outline of their footprints; but there were no clouds of dust above the airless surface. Most of the finer dust has evidently been pounded down into the Moon by the churning of meteorites.

FIG. 2.4 Moonwalk Charles Duke strolls across the lunar surface during the *Apollo 16* mission in April 1972. Small impacting particles have sandblasted the lunar surface, producing smoothed, undulating layers of fine dust and rounding the surfaces of lunar rocks. Larger meteorites have pounded and churned the surface, producing a layer of rocky debris. (Courtesy of NASA.)

In all, twelve humans have roamed the surface of the Moon and brought back nearly half a ton, or 500 kilograms, of rocks. The lunar samples contain no water and demonstrate that there is no life on the Moon – and that there apparently never was any. Radioactive dating of the rocks indicates that most of the features we now see on the Moon have been there for more than 3 billion years. And as the result of the *Apollo* missions, we are pretty sure that the Moon originated long ago during a collision of a Mars-sized body with the young Earth; the giant impact dislodged material that would become the Moon that we know.

But to be honest, the main motivation in the race to the Moon was not scientific. It was the Cold War rivalry between the United States and the Soviet Union. And the American achievement was a spectacular political triumph. The dissipation of the Soviet Union's lead in space tarnished the image of Soviet competence and diminished their status in world affairs. In contrast, landing men on the Moon and conquering the frontier of space taught the American people that nothing is impossible if they set their sights high enough; with resolve and will power you can accomplish anything, especially in a democratic nation that stresses individual freedom. And it was surely a contributing factor to the idea that success can result from technological superiority.

No one has walked on the Moon since 1972, but the Americans might soar into outer space again. In 2004 President George W. Bush announced plans to send astronauts back to the Moon and later Mars. Summoning the spirit of discovery, exemplified by the Lewis and Clark exploration of the vast American West two centuries before, Bush called on the human need to venture into space, "for the same reason we were once drawn into unknown lands and across the open sea,"[23] urging NASA to return people to the Moon no later than 2020.

Technology can now be used that wasn't even imagined when humans first walked on the Moon. We might eventually establish a permanent human settlement there, using it as a base for the exploration of Mars. And since the Moon has no significant atmosphere, some promising astronomy could be accomplished from it. Nevertheless, science is not likely to be the main impetus this time around, any more than it was during the first visits.

But in the meantime, the voyage to the Moon had become the stepping-stone to outer space, resembling the first, tentative steps of a child testing the water before learning to swim. It paved the way for an ongoing, close-up exploration of the planets and their satellites, which have been mapped and surveyed with a detail surpassing that of most countries on our home planet, Earth.

2.3 THE CLOSE-UP VIEW FROM SPACE

The initial exploration of the Moon established a blueprint for missions to all the major planets. Spacecraft flew by them to provide the initial reconnaissance. This was followed by a more detailed exploration with orbiting spacecraft, which can circle a planet many times and map out its terrain. Like the explorers of new territories on Earth, the orbiters were sent to reconnoiter, to get the lay of the land, and to disclose possible dangers awaiting future visits. The next step involved landers that explored the surfaces, and probes that plunged into the atmospheres. All of these spacecraft have fundamentally altered our perception of the Solar System, providing images of landscapes that cannot be directly seen from Earth. They have forever changed our view of the planets and their moons, including a growing awareness of their similarity.

Signs of past volcanism are found throughout the Solar System. In the early 1990s, the *Magellan* spacecraft orbited the planet Venus, using its radar to penetrate the clouds that perpetually hide the planet's surface from view. The radar images indicate that lava smoothed out the entire surface of Venus about 750 million years ago, and that tens of thousands of volcanoes now pepper the planet's flattened surface (Fig. 2.5).

No one expected volcanoes on Jupiter's innermost large moon, Io, which is a relatively small rocky satellite the size of our Moon and located in the frozen outskirts of the Solar System. So everyone was amazed when the *Voyager 1* spacecraft discovered active volcanoes on Io in 1979, which were still erupting during the *Galileo* spacecraft's rendezvous from 1995 to 2003. Io is the most volcanically active body known, with volcanoes that are turning the satellite inside out and transforming its surface before our very eyes (Fig. 2.6). Changing tidal forces, induced in the satellite by nearby massive

FIG. 2.5 Maat Mons This three-dimensional perspective of Maat Mons on Venus was obtained using radar data taken from the *Magellan* spacecraft in October 1991. The volcano is 8 kilometers high, the second highest peak on the planet. Fresh, dark lava extends down and across the foreground, perhaps flowing from a relatively recent eruption. Maat Mons is a giant shield volcano similar in size and shape to the big island of Hawaii. *Maat* is the name of the ancient Egyptian goddess of truth and justice, and *Mons* is the Latin term for "mountain". (Courtesy of NASA and JPL.)

FIG. 2.6 **Volcanic deposits on Jupiter's satellite Io** Massive eruptions continually disfigure the surface of Jupiter's volcanically active moon Io. A bright red ring surrounds the volcano Pele, marking the site of sulfur compounds deposited by its volcanic plumes. A dark circular area intersects the upper-right part of the red ring and surrounds another volcanic center named Pillan Patera. Deposits of sulfur dioxide frost appear white and gray in this image, while other sulfurous materials probably cause the yellow and brown shades. Pele is the Hawaiian goddess of the volcano, and Pillan Patera is named for the Araucanian thunder, fire and volcano god. This image was taken on 19 December 1997 from the *Galileo* spacecraft. (Courtesy of NASA and JPL.)

Jupiter, squeeze Io's rocky interior in and out, making it molten inside and producing the volcanoes.

Water is also ubiquitous in the Solar System, and because water is a key prerequisite for life, it provides evidence that places like Mars may have been habitable at sometime during their history.

Although Mars is now a cold and desolate world (Fig. 2.7), with vast quantities of water frozen below its surface, global images taken in the 1970s from the *Mariner 9* and

FIG. 2.7 Surface of Mars The Martian surface in western Chryse Planitia, as viewed from the *Viking 1* lander on 3 August 1976. Wind-blown dust clings to eroded rocks, creates dust drifts, and enters the sky. Mars is a frozen, desolate planet whose gently rolling landscape resembles sand dunes on a rocky terrestrial desert. But the dune drifts on Mars are composed of fine dusty material and the Martian dust drifts persist for much longer than sand dunes on Earth. These drifts were little changed during the six years they were observed from *Viking 1*. (Courtesy of NASA and JPL.)

Viking 1 and *2* spacecraft revealed dry riverbeds, deep winding channels, and streamlined, washed-out landforms, providing compelling evidence for abundant liquid water on Mars in the distant past. In the planet's early history, rivers ran across its surface and powerful floods coursed down its valleys, emptying into the russet plains and perhaps forming ancient lakes or seas.

In recent years, crisp, sharp images from the *Mars Global Surveyor* have recorded layered deposits that might have been laid down in former lakes and shallow seas. The intrepid spacecraft has even captured evidence for liquid water that may have been seeping out of the walls of Martian canyons and craters in recent times, creating small gullies and depositing the debris in fan-like deltas. Moreover, the composition, minerals and structures found within rocks sampled by the Mars rover *Opportunity* in 2004 provide direct evidence for sustained periods of flowing water sometime during the red planet's history.

Jupiter's satellite Europa is a cracked ball of water ice (Fig. 2.8). Deep long fractures run like veins through the icy covering, apparently filled by the upwelling of dirty

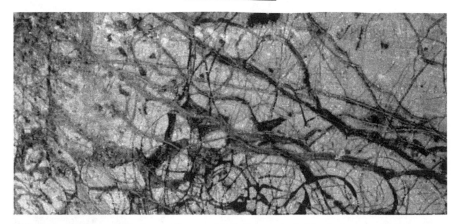

FIG. 2.8 Broken water ice on Europa Dark linear, crack-like features extend across most of Jupiter's moon Europa. The fractures may have been filled by liquid water or dirty slush gushing out from a global ocean in the satellite's interior. The long cracks were most likely caused by tides raised on Europa by the gravitational pull of Jupiter, and the absence of craters suggests that the crust of water ice is relatively young. This image, about 770 kilometers wide, was taken from the *Galileo* spacecraft on 27 June 1996. (Courtesy of NASA and JPL.)

liquid water, and large blocks of water ice move at slow glacier pace across the surface. An ocean of liquid water apparently exists at shallow depths beneath the frozen surface, lubricating overlying ice rafts and oozing out into cracks in its icy covering. Since Europa is farther from Jupiter than Io, there is less heat generated by tidal flexing, but it may still be enough to liquefy Europa's internal ice.

The austerely beautiful rings of Saturn are composed of innumerable particles of water ice (Fig. 2.9). Modern spacecraft, such as *Voyager 1* and *2* in 1980 – 81 and *Cassini* in 2004 have shown that small nearby satellites sharpen the edges of the fabled rings, and produce gaps, numerous ringlets, and waves in them.

And comets are nothing more than relatively small balls of dirty ice, about the size of Paris or Manhattan. When a comet comes near the Sun, the solar heat vaporizes the comet ice, producing stately comet tails. As it moves away from the Sun, a comet receives less solar heat, becoming cold and inert and fading into darkness.

Three brave satellites have penetrated the fluorescing comet gas, to catch a glimpse of the dark, tar-like crust covering the comet nucleus, with bright jets that spew gas and dust from fissures on its sunlit side (Fig. 2.10).

Thus, modern spacecraft have transformed our vision of the planets and moons with fantastic images of previously unseen worlds. And now telescopes in space and on the ground have been used to show that planets are not unique to our own Solar System. They have found that young stars are enveloped by swirling disks of gas and dust, which are expected to coalesce into planets, that Earth-sized planets orbit at least one pulsar, and that numerous Jupiter- and Neptune-sized globes are revolving around ordinary Sun-like stars.

FIG. 2.9   **Saturn's rings of ice** The narrow angle camera aboard the *Cassini* spacecraft took this image from beneath Saturn's ring plane on 21 June 2004. The brightest part of the ring system, extending from the upper right to the lower left, is the central B ring. It is separated from the outermost A ring by the wide, dark Cassini Division, discovered in 1675 by the Italian-born French astronomer Giovanni (Gian) Domenico (Jean Dominique) Cassini (1625–1712). Below the B ring, closer to the planet, is the C ring. All three rings are composed of innumerable particles of water ice. The different shades of the rings are attributed to different amounts of contamination by other materials such as rock or carbon compounds. When viewed close up, the broad icy rings break up into thousands of individual wave-like ringlets. (Courtesy of NASA, JPL and SSI.)

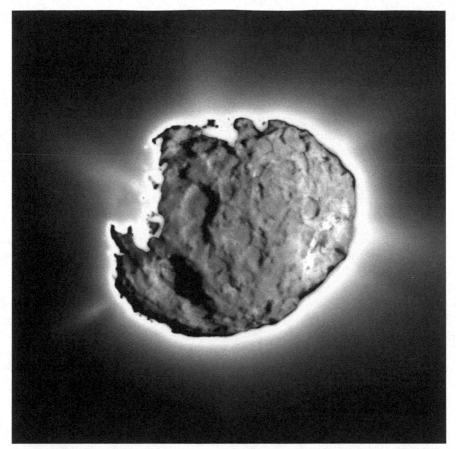

FIG. 2.10 **Comet nucleus** The black, solid nucleus of Comet Wild 2 is a dirty ball of water ice about five kilometers across, with several large depressed regions on its dark surface. It is silhouetted against bright jets of water and dust that stream outward into space, from the intensely active surface. This is a composite of surface and jet images taken from the *Stardust* spacecraft on 2 January 2004. *Stardust* also collected comet dust particles, which were returned to Earth in January 2006. Since the comet is a fairly recent arrival to the inner Solar System, its material has probably not been altered significantly by the heat of the Sun. Terrestrial analysis of the comet samples will therefore provide clues to the earliest history of our Solar System about 4.6 billion years ago. (Courtesy of NASA and JPL.)

2.4 PLANETS AROUND OTHER STARS

The Plurality of Worlds

The ancient Greeks imagined that all matter consists of tiny moving particles, both indivisible and invisible, which they called atoms, and that all material things can be created by the coming together of a sufficient number of atoms. Around 300 BC, the Greek philosopher Epicurus proposed that the chance conglomerations of innumerable atoms, in an infinite Universe, should result in the formation of a multitude of unseen worlds, each with an Earth-like planet at the center. And his disciple, Metrodorus of Chios, wrote "it would be strange if a single ear of corn grew in a large plain or were there one world in the infinite."[24]

In 55 BC, the Roman poet Lucretius also wrote about the plurality of worlds within a Universe without end:

> Granted, then, that empty space extends without limit in every direction and that seeds innumerable in number are rushing on countless courses through an unfathomable Universe under the impulse of perpetual motion, it is in the highest degree unlikely that this Earth and sky is the only one to have been created and that all those particles outside are accomplishing nothing. This follows from the fact that our world has been made by nature through the spontaneous and casual collision and the multifarious, accidental, random and purposeless congregation and coalescence of atoms.[25]

Astronomers eventually showed that the planets in our Solar System are revolving around the Sun, and that the Sun is itself one of innumerable stars. These discoveries opened up the possibility that there might be planets orbiting other stars, some possibly inhabited. Such a belief dates back at least as far as the Italian philosopher and priest Giordano Bruno, who reasoned in 1584 that these other planets should be orbiting bright stars, and would "remain invisible to us because they are much smaller and non-luminous."[26]

Bruno spent the last eight years of his life in the prisons of the Inquisition. He was eventually tried by the Catholic Church, bound to a stake, and burned alive in Rome in 1600, perhaps more for his heretical religious views, such as his doubts about the Immaculate Conception, the Holy Trinity, and Christ's divinity, then for his belief in an infinite Universe filled with countless habitable planets circling other stars.

One approach to these speculations about the plurality of worlds is to consider the origin of our Solar System.

The Nebular Hypothesis and Planet-Forming Disks

Any successful theory for the origin of the planets in our Solar System must account for the regular arrangement of their orbits. They all move in a narrow band across the sky, the zodiac, which implies that the orbits lie nearly in a plane, and they all orbit the Sun in the same direction that our star rotates. These orbital paths are nearly circular, and the equator of the Sun's rotation nearly coincides with the plane of the planetary orbits. This highly regular pattern, which cannot be accidental, is a natural consequence of the nebular hypothesis, in which the Sun and planets formed together out of a single collapsing, rotating cloud of interstellar gas and dust, called the Solar Nebula.

The basic idea was introduced by the German philosopher Immanuel Kant in his book *Allgemeine Naturgeschichte und Theorie des Himmels,* or *Universal Natural History and the Theory of the Heavens,* published in 1755. Kant pictured an early Universe filled with thin gas that collected into dense, rotating gaseous clumps. One of these primordial concentrations was the spinning solar nebula. Attracted by its own gravity, the nebula fell in on itself, getting denser and denser, until the middle became so packed, so tight and hot, that the Sun began to shine. Meanwhile, the rotation spun the surrounding material out into a flattened disk revolving about the central, contracting proto-Sun (Fig. 2.11). The planets formed from swirling condensations in this circumstellar material, which explains qualitatively why all the planets revolve in the plane that coincides with the equator of the rotating Sun.

Astronomers have discovered flattened, rotating disks encircling other stars, suggesting that the nebular hypothesis applies to them. This material is expected to coalesce into full-blown planets if it hasn't already done so.

Instruments aboard the *InfraRed Astronomical Satellite*, abbreviated *IRAS* and a collaborative project of the United States, Britain and Holland, unexpectedly obtained the first evidence for such planet-forming disks in 1983. Using technology pioneered by the military to detect the infrared heat of the enemy, the satellite was designed to detect cosmic infrared radiation, which is mainly inaccessible from the ground owing to absorption in the atmosphere. Because of their comparatively low temperature, dust particles emit most of their radiation at infrared wavelengths, while radiating no detectable visible light. It is the other way around for the hot stars, which shine brightly in visible light and emit relatively little infrared.

FIG. 2.11 Formation of the Solar System An artist's impression of the nebular hypothesis, in which the Sun and planets were formed at the same time during the collapse of a rotating interstellar cloud of gas and dust that is called the solar nebula. The center collapsed to ignite the nuclear fires of the nascent Sun, while the surrounding material was whirled into a spinning disk where the planets coalesced. (Courtesy of Helmut K. Wimmer, Hayden Planetarium, American Museum of Natural History.)

The *IRAS* instruments found the bright infrared glow of dusty clouds, disks and rings circling bright, massive stars such as the brilliant blue-white Vega, as well as less-massive, solar-type stars. In fact, the youngest nearby stars are usually found embedded in the dense clouds of interstellar gas and dust that spawned them. And the *Spitzer Space Telescope* has recently used its powerful infrared vision to detect hundreds of stars with excess infrared radiation, suggesting that they harbor planet-forming disks.

The *Hubble Space Telescope*, abbreviated *HST*, has discovered flattened disks of dust swirling around at least half the young stars in the Orion Nebula, shining in reflected visible light. The high-resolution and sensitivity of the *HST* have also been used to obtain detailed images of dusty, planet-forming disks surrounding other Sun-like stars, providing insights to the beginnings of our Solar System (Fig. 2.12).

Individual planets shine by reflecting light that is much fainter than the light of the star that illuminates them. The visible light reflected by Jupiter is, for example, about a billion, or 10^9, times dimmer than the light emitted by the Sun, and that reflected by Earth is 10 billion times fainter. As a result, distant planets are too small and too faint to be seen directly in the bright glare of their nearby star. Their presence has only recently been inferred from their miniscule gravitational effects on the motions of the star they revolve around.

Pulsar Planets

In 1992, two American radio astronomers, Aleksander Wolszczan and Dale A. Frail, reported the first discovery of planets outside the Solar System, unexpectedly orbiting a

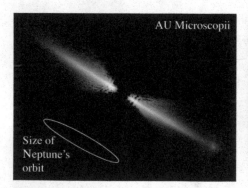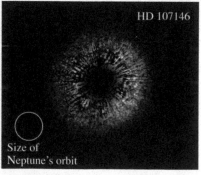

FIG. 2.12 Dusty disks around Sun-like stars Instruments aboard the *Hubble Space Telescope* have obtained these images of the visible starlight reflected from thick disks of dust around two young stars that might still be in the process of forming planets. Viewed nearly face on, the debris disk surrounding the Sun-like star known as HD 107146 (*right*) has an empty large enough to contain the orbits of the planets in our Solar System. Seen edge-on, the dust disk around the reddish dwarf star known as AU Microscopii (*left*) has a similar cleared-out space in the middle. HD 107146 is 88 light-years away, and is thought to be between 50 million and 250 million years old, while AU Microscopii is located 32 light-years away and is estimated to be just 12 million years old. (Courtesy of NASA, ESA, STScI, JPL, David Ardila–JHU (*right*), and John Krist–STScI/JPL (*left*).)

pulsar, designated PSR 1257+12 for its position in the sky. As with other pulsars, it is a tiny, superdense neutron star that emits precisely periodic radio radiation as it rotates.

Wolszczan wasn't looking for planets. He was using the giant radio telescope at Arecibo, Puerto Rico to search for pulsars that spin very rapidly, at the rate of several hundred times a second. Since the telescope is always in high demand, the search only became possible when it was shut down for repairs and pointing in just one direction in the sky. As luck would have it, a previously unknown, fast-spinning pulsar happened to pass through the immobile antenna beam.

Once the telescope was repaired, Wolszczan used large computers and comparisons to atomic clocks to accumulate and time the arrival of millions and then billions of the rapid, uniform pulses, obtaining a very accurate repetition period. The best results indicated that the pulsar was spinning once every 0.0062 seconds, or 6.2 milliseconds, rotating on it axis 161 times every second. But there seemed to be something wrong with the data, for the repeating pulses did not match a single, well-defined periodicity, and the computer could not predict exactly when each pulse would arrive at Earth.

The position of the pulsar could be slightly in error, causing a mismatch in the computer analysis of the pulsar timing, which corrects for the motion of the Earth towards any point in space. So Dale Frail used the Very Large Array in New Mexico to obtain an accurate location. With the new position, the timing data made sense. The pulse arrival times were being affected by at least one unseen planet.

When a planet tugs the pulsar toward us, its pulses are bunched up and more pulses are detected every second. And when the pulsar is pulled away, the pulses spread out with fewer pulses received every second. It is this periodic stretching and compression of the pulsar radio waves that produced minute, periodic changes in the measured repetition period of the star's otherwise regular pulses of radiation.

The effect is entirely analogous to the well-known Doppler effect, discovered by the Austrian physicist Christian Doppler. In 1892 he described how sound depends on the relative motion of the source and listener. If the source is moving toward us, the motion compresses the sound waves, pushing them at us and shortening their wavelength. Sound emitted at a given wavelength then arrives at a shorter wavelength than that emitted by a stationary source; more crests strike the ear each second and a higher frequency or pitch is heard. The sound waves of a receding source are pulled away from us and stretched out to longer wavelengths, with a lower pitch than would be emitted by a non-moving source.

Just as a source of sound can vary in pitch or wavelength, depending on its motion, the wavelength of electromagnetic radiation shifts when the emitting source moves with respect to the observer. If the motion is toward the observer, the Doppler shift is to shorter wavelengths, and when the motion is away the wavelength becomes longer. The amount of the wavelength shift can be used to infer the velocity of motion along the line of sight, known as the radial velocity.

As it turned out, at least two planets were affecting the pulsar radio waves, each with a mass comparable to that of the Earth, nudging the pulsar slightly toward and away from us with their gravitational pull.

It was a momentous occasion, for the first planets had been found outside our Solar System. But the response of the astronomical community was at best lukewarm, for the planets are orbiting the wrong kind of star. Without any thermonuclear fuel to make it shine, the pulsar PSR 1257 + 12 is a cold, dark star, emitting no light to warm the newfound planets. In addition, the intense, spinning magnetic fields of the pulsar must accelerate and expel lethal high-energy particles and radiation.

So astronomers were disappointed. They had hoped to discover planets around a perfectly ordinary star like the Sun, whose steady light and heat would at least be compatible with the notion of extraterrestrial life.

The First Unseen Planets Around Ordinary Stars

If any planet in our Solar System were placed in orbit around any other star, it would vanish from sight, lost in the star's glare. The presence of the unseen planet has to be instead deduced by indirect means, by recording the way its gravity pulls the star around. The more massive the planet, and the closer it is to the star, the stronger the planet's gravitational pull on the star and the more the planet perturbs it.

Like two linked, whirling dancers, the planet and star tumble through space, pulling each other in circles. They both orbit a common center of mass where their gravitational forces are equal, somewhat like the equilibrium point of a seesaw, where the forces of two people balance. This fulcrum is closest to the heavier person, or to the massive star in the stellar case. So the star moves in a much smaller circle, a miniature version of the planet's larger path.

To detect this tumbling motion, astronomers had to look for subtle changes in starlight as the unseen planet tugged on the star, pulling it first toward the Earth and then away, causing a periodic shift of the stellar radiation to shorter and then longer wavelengths (Fig. 2.13). But the variation in the wavelength of stellar spectral features, called lines, is exceedingly small. Massive Jupiter, for example, makes the Sun wobble at a speed of only about 12 meters per second. To detect the Doppler effect of a star moving with this speed, astronomers would have to measure the wavelengths with an unheard of accuracy of at least one part in 100 million.

So the effect could only be detected if very sensitive spectrographs were constructed to precisely spread out the light rays. The enhanced light-collecting powers of electronic charge-coupled detectors were also needed to record the dispersed starlight. And since no single line shift is significant enough to be seen, computer software had to be written to add up all the star's spectral lines, which shift together, combining them over and over again at all possible regularities, or orbital periods, with continued comparison to non-moving laboratory spectral lines.

It took decades for astronomers to develop the complex and precise instruments. Then, in the 1990s, the time was ripe, and two Swiss astronomers from the Geneva Observatory in Switzerland, Michel Mayor and Didier Queloz, accomplished the seemingly impossible. In April 1994 they attached a new, exquisitely precise, computerized spectrograph to the 1.93-meter (76-inch) telescope at the Observatoire de Haute-Provence in the south of France, and within a year and a half they had found the first planet that orbits an ordinary star, the faintly visible, Sun-like star 51 Pegasi, only 48 light-years away from Earth in the constellation Pegasus, the Winged Horse.

Hints of the planet were found a year before the announcement of its existence, but the Swiss team had to be very careful. There had been many false planetary

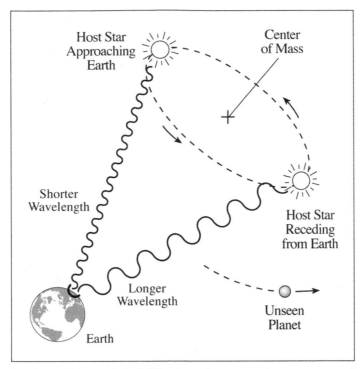

FIG. 2.13 Starlight shift reveals invisible planet An unseen planet exerts a gravitational force on its visible host star. This force tugs the star in a circular or oval path, which mirrors in miniature the planet's orbit. As the star moves through space, it approaches and recedes from Earth, changing the wavelength of the starlight seen from Earth through the Doppler effect. When the planet pulls the star toward us, its light waves pile up in front of it slightly, shortening or "blueshifting" the wavelength we detect. When the planet pulls the star away from us, we detect light waves that are stretched or redshifted. During successive planet orbits, the star's spectral lines are periodically shortened and lengthened, revealing the presence of the planet orbiting the star, even though we cannot see the planet directly.

discoveries before, and they seemed to have found a giant planet with an unexpectedly short orbital period of just 4.23 days. By way of comparison, giant Jupiter orbits the Sun once every 11.86 *years*. Observations were stopped in March 1995 because the star moved too close to the Sun to be seen, and renewed in the first week of July when the two astronomers returned with their families. Armed with a precise prediction of what the spectrograph would show if the unseen planet really existed, they pointed the telescope at 51 Pegasi, and saw exactly what they had hoped for. As Mayor described it, the occasion happened like a dream, a spiritual moment.[27]

Mayor and Queloz submitted a discovery paper to *Nature* magazine the following month, and announced it at a professional meeting in Firenze, Italy on 6 October 1995.

They had detected the back-and-forth Doppler shift of the star's light with a regular 4.23-day period, measured by a periodic excursion of the star's radial velocity of up to 50 meters per second (Fig. 2.14). To produce such a quick and relatively pronounced wobble, the newfound planet had to be large, with a mass comparable to that of Jupiter, which is 318 times heftier than Earth, and moving in a tight, close orbit around 51 Pegasi.

Planets that are closer to a star move around it with greater speed and take less time to complete an orbit, all in accordance with Kepler's third law. Thus, the Earth

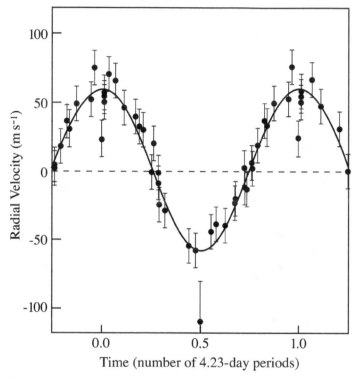

FIG. 2.14   Unseen planet orbits the star 51 Pegasi Discovery data for the first planet found orbiting a normal star other than the Sun. The giant, unseen planet is a companion of the solar-type star 51 Pegasi, located 50 light-years away. The radial velocity of the star, in units of meters per second, has been measured from the Doppler shift of the star's spectral lines. The velocity exhibits a sinusoidal variation with a 4.23-day period, caused by the invisible planetary companion that orbits 51 Pegasi with this period. The observational data (*solid dots*) are fit with the solid line, whose amplitude implies that the mass of the companion is roughly 0.46 times the mass of Jupiter. The 4.23-day period indicates that the unseen planet is orbiting 51 Pegasi at a distance of 0.05 AU, where 1.00 AU is the mean distance between the Earth and the Sun. [Adapted from Michael Mayor and Didier Queloz, "A Jupiter-mass companion to a solar-type star", *Nature* **378**, 355–359 (1995).]

takes a year or 365 days to travel once around the Sun at a mean distance of one astronomical unit, or 1 AU, while Mercury, the closest planet to the Sun, orbits our star with a period of 88 days at 0.387 AU. A short orbital period of only 4.23 days meant that the newfound planet is located at a distance of just 0.05 AU from its parent star, or about one-eighth the distance between Mercury and the Sun. Thus, a completely unanticipated planet had been found, rivaling Jupiter in size and revolving around 51 Pegasi in an orbit smaller than Mercury. No one expected a giant planet to be revolving so close to its star.

Less than two weeks after the announcement of a giant planet circling 51 Pegasi, two American astronomers Geoffrey W. Marcy and R. Paul Butler confirmed the result using the 3-meter (120-inch) telescope at Lick Observatory near Santa Cruz, California. On 17 October 1995, Marcy and Butler issued a press release containing the confirmation, which hit the front-page headlines of newspapers throughout the world. And now that they knew that giant planets could revolve unexpectedly near a star, with short orbital periods, they used powerful computers to re-examine their data accumulated during previous years, announcing in January 1996 the discovery of two more Jupiter-sized companions of Sun-like stars.

These were astounding discoveries. In just a few months, astronomers had detected the first planets circling ordinary stars just like our Sun. Other worlds were no longer limited to philosophical musings, scientific speculations, or artists' imaginations (Fig. 2.15). After two millennia, a long dream has come true. We can now look up at the night sky and say that there are definitely invisible planets out there, orbiting perfectly ordinary stars shining brightly in the night sky.

Hundreds of New Worlds Circling Other Stars

Scientists had spent decades looking for giant planets far from their central star, only to find that they are easy to detect once you look close in. After scientists realized that a large planet could be so near to its star, they knew where and how to look. And by monitoring thousands of nearby Sun-like stars for years, American and European teams have found more than one hundred massive, Jupiter-sized planets revolving about other stars.

Some of the newfound worlds travel in circular orbits, like those in the Solar System, but much closer to their parent stars than Mercury is to the Sun. Dubbed "hot Jupiters" because of their proximity to the intense stellar heat, their temperatures can soar to more than 1000 kelvin, far hotter than the surface of any planet in our Solar System. Other newfound planets follow eccentric, oval-shaped orbits that deviate from a circular path, so they venture both near and far from their stars.

Most of these worlds have been discovered by the wobble they create in the motion of their parent star, but some of them were discovered when they passed in front of the star, causing it to dim, or blink. Such a transit can only be seen if the orbit of the distant planet crosses the line of sight from Earth, blocking a tiny fraction of the star's observed light and causing it to periodically dim, over and over again during the planet's endless journey around the star. The *Spitzer Space Telescope* has even been used to measure the warm infrared glow of two of the hot giant planets that were previously discovered by the transit method, showing that they have temperatures of at least 1,000 kelvin.

FIG. 2.15   Hair followed by two planets The Spanish artist Joan
Miró (1893–1983) painted this work in 1968, showing two planets
that seem to move about a remote star, another Sun. It is a poetic and
prophetic vision of planets orbiting nearby stars, discovered nearly
half a century later. Oil on canvas, 1.95 × 1.30 m (Courtesy of the
Pierre Matisse Gallery, New York.)

As the years passed and observations continued to accumulate, it became possible
to detect a few giant planets with far out orbits and long orbital periods, comparable
to Jupiter's 5.2 AU and 11.86-year orbit. Multi-planet systems have also been found
as the result of longer and improved observations, such as the three giant worlds that
circle the solar-type star Upsilon Andromedae, located 44 light-years away and visible
to the unaided eye (Fig. 2.16).

FIG. 2.16   Planets orbit the star Upsilon Andromedae Astronomers have detected three large planets orbiting the solar-type star Upsilon Andromedae, located at a distance of 44 light-years. The first, innermost, planet, denoted by the letter *b*, was discovered in 1996. It

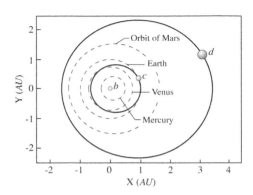

orbits the star with a period of 4.617 days at a distance of 0.059 AU, much closer than Mercury orbits the Sun. Several years of additional observations revealed two distant planets, designated *c* and *d*, whose orbital periodicities of 241.5 days and 1,284 days have been inferred from the plot of the radial velocity residuals with the 4.617-day periodicity removed. These two planets revolve about their star in eccentric ovals (*solid lines*), one moving between distances comparable to the distance of Venus and the Earth from the Sun and the other at distance further than the orbit of Mars around the Sun (*dashed lines*). All three planets are heavy giants, with estimated masses of 0.69, 1.19 and 3.75 times the mass of Jupiter, for planets *b, c* and *d,* respectively. (Courtesy of Geoffrey Marcy and colleagues.)

And this is just a beginning. More than ten planetary systems have been discovered orbiting stars other than the Sun, each containing two or more giant planets, about as massive as Jupiter or more, and these new worlds all circle their parent stars closer than Jupiter's orbital distance from the Sun. Three smaller planets have also been detected, which are just 14 to 20 times more massive than the Earth. So they are comparable to Uranus and Neptune in mass, which weigh in at 14.5 and 17.1 Earth masses, respectively, but significantly less massive than Jupiter. These discoveries have taken astronomers a step closer to finding even smaller Earth-sized planets that might orbit nearby stars at distances where life could exist.

Searching for Habitable Planets

From a human perspective, the most interesting planets would be those as small as the Earth, in circular orbits at just the right distance from the heat of a Sun-like star to provide a haven for life. But our instruments do not yet have the precision needed to detect such an Earth-sized world outside our Solar System, either near or far from an ordinary star. They produce too small a gravitational perturbation of their star to be detected by the Doppler method with existing technology.

The transit method might be used to detect Earth-sized planets. The orbital size can be calculated from the period of the repeated eclipse, and the planet's temperature estimated. This information would tell us if the planet resides within a habitable zone, the range of distances from a star for which liquid water can exist. At closer distances, the water would all be boiled away, and at more remote distances it would be frozen solid. In addition, the amount of starlight dimming provides a measurement of the planet's size and probable mass. Two spacecraft missions are now being planned to observe such planetary transits from space.

In the next decade, light from two or more separate telescopes will also be combined to yield incredibly sharp views of planetary systems, first on the ground and then from space. The technique is known as interferometry because it analyzes how the light waves detected at the telescopes interfere once they are added together. Unencumbered by our atmosphere, one such project, called the *Space Interferometry Mission* or *SIM* for short, will be able to detect planets as small as a few Earth masses revolving around a number of nearby stars in the habitable zone where liquid water can exist.

But these are the dreams of our future, and at the moment we have no direct evidence for any life elsewhere than on Earth. So we now turn to the larger fabric of the Universe, to the stars and galaxies, whose motions give them shape and form.

Chapter Three

Motion, Content and Form

3.1 STOP MOVING AND YOU'RE DEAD

Look at us, with our fast-paced, distracted and impatient lives, accelerated by the speed of our cars and split-second access to unlimited information on our computers. Parents rush to work or to take children to school, jet airplanes carry busy people to important meetings, and even the birds seem to be hurrying though on their way to some other place. And nothing stays still down inside our bodies, where the blood cycles through our veins and molecules move and vibrate in endless interaction.

It seems that you can't exist without moving; otherwise gravity or something else will pull you down. Or perhaps we have to move quickly to keep from falling, or because we don't know what's going to happen. Stop moving and it's all over, or as Bob Dylan sang, "you better start swimming, or you'll sink like a stone."[28]

Motion combines space and time to produce a changing perspective, enhancing our vision and making it easier to see some things. Our eyes catch the motion of a windblown leaf as it marks out a trajectory against the stationary background, or the luminous flash of a kingfisher bird as it dive-bombs into a lake for lunch. And when motion ceases, it can seem like time stopping, as when a rocking boat docks and one disembarks to solid land.

The Russian painter Wassily Kandinsky was a master at using line, form and color to depict movement, an unfolding in space and time. He knew that our gaze moves from side to side and from near to far when absorbing the elements of a picture. So he would capture this motion, and extend a painting's depth and breath with an arrangement of lines, colored circles and other geometrical designs that the eye follows (Fig. 3.1), often placing them within a stationary black background, like the empty night sky (Fig. 3.2).

Motion can be introduced into a picture by painting different parts in various colors, which can signify approach or recession. A red element seems to advance toward our eyes, while a blue one seems to move away. And perhaps that is not so surprising. We are naturally drawn toward the warmth of a red fire on a cold night, and accustomed to glancing up and out into the blue sky. When the two colors are placed near each other in a painting, they can produce resonating in-and-out vibrations as the eye moves back and forth between them.

FIG. 3.1 **Radiating lines** Our restless eyes follow the lines, triangular forms and colors of this 1927 painting by the Russian artist Wassily Kandinsky (1866–1944), creating a flow, direction, and pictorial movement, like the unfolding of a dance in space or of music in time. The lines radiate from just one side of a large object, set against a background of smaller, immobile red or yellow disks. The triangular shapes in the lower left resemble a spacecraft propelled upward by the thrust of engine exhaust in the opposite direction. Oil on canvas, 0.75 × 1.0 m. (Courtesy of the Solomon R. Guggenheim Museum, New York City, New York.)

The American sculptor Alexander Calder added a fourth, temporal dimension in his moving sculptures, called mobiles. They are made from hinged pieces of wire and attached forms, suspended from a fixed point, weighted and delicately balanced so they move in the slightest breeze. Unpredictable and ever changing, a mobile seems alive, in perpetual motion without beginning or end. Its individual elements move back and forth and around in closed paths, like the leaves of a tree quivering in the wind before a storm.

FIG. 3.2 Movement 1 All resemblance of the physical world
has been eliminated in this work, executed in 1935 by the Russian
artist Wassily Kandinsky (1866–1944). The objects seem to move,
imparting a cosmic energy to the distant Universe. Oil and tempura
on canvas 1.16 × 0.89 m. (Formerly collection of Nina Kandinsky,
Paris, now located in Moscow).

Astronomers encapsulate space and time when watching the moving Moon, planets or stars. The Moon moves across the night sky at a stately pace, the Earth and every other planet spins on its axis and whirls about the Sun, and the Sun and other stars dart here and there, like bees in a swarm, while revolving at much faster speeds about a massive center.

So the apparently serene Cosmos, which is frozen into a single glance at the night sky, is an illusion, for there is nothing in it that is completely at rest. Everything that exists, from atoms to planets and stars to galaxies, moves through space, all surely going somewhere. It is this motion that shapes the Universe, giving it form, structure and texture.

3.2 THE PLANETS MOVE

The Wanderers

Our remote ancestors looked up on any dark, moonless night, and saw thousands of stars. They identified the brightest by name and noticed patterns, now called constellations, amongst groups of them. These permanent stellar beacons are always there, firmly rooted in the dark night sky, and the constellations remain seemingly unchanged over the eons. That is why astronomers say that the stars are fixed, apparently cemented in place within the night sky.

So ancient astronomers naturally focused attention on a few wandering objects. One of them is our Moon, whose changing appearance and repetitive phases clock our monthly cycle. Her bright, changing and moving face has captured the imagination of poets for centuries, including John Milton's:

> To behold the wandering Moon,
> Riding now her highest noon,
> Like one that has been led astray
> Through the heav'n's wide pathless way.[29]

Or Samuel Coleridge's:

> The moving Moon went up the sky,
> And nowhere did abide.
> Softly she was going up,
> And a star or two beside.[30]

And the Sun also rises and sets each day, and moves across the sky with a yearly rhythm, never rising at precisely the same spot on the horizon each day. Instead, the location of sunrise drifts back and forth along the horizon, while climbing higher or lower in the sky, in an annual cycle. The moving Sun and Moon therefore provided the first clocks, marking out our daily lives, the months, and the years.

Altogether, the ancients knew of seven objects that moved against the unchanging stellar background. Ranked in order of greatest apparent brightness, they are the Sun, the Moon, Venus, Jupiter, Saturn, Mercury and Mars. The Greeks called them *planetes*, their word for "wanderers;" and we still use the designation planets for all but the Sun and Moon.

According to Plato, writing around 380 BC, the simplest and purest sort of motion is circular, so circles ought to describe the visible paths of the moving planets. After all, a wheel moves so well because it is round, without rough, sharp edges to get in the way. The circle also has no beginning or end, seemingly appropriate for describing the endless motion of the heavenly wanderers. And the central Earth would be separated from the heavens, like a magician who draws a boundary circle around himself to seal off the region in which magical powers are brought into play.

Following Plato's suggestion, astronomers spent centuries trying to discover those uniform, perfectly regular, circular motions that would "save the appearances" presented by the planets.[31] They supposed that the Earth stood still, an immobile globe at the center of it all. The imaginary celestial sphere of fixed stars wheeled around the central, stationary Earth once every day, with uniform circular motion and perfect regularity, night after night and year after year. Such a celestial sphere would also explain why people located at different places on Earth invariably saw just half of all the stellar heavens, and why travelers to new and distant lands would see new stars as well as new people.

The Sun, Moon and planets were supposed to be carried on concentric, transparent crystalline spheres, which revolved around the stationary Earth, but their hypothetical spinning motion contradicted the observed planetary motions. So the Egyptian astronomer, Claudius Ptolemy, shifted the Earth from the exact center of the Universe, by just a small amount, and described the planetary appearances with an intricate system of circles moving on other circles, like the gears of some fantastic cosmic machine. Ptolemy was thereby able to reproduce the apparent motions of the planets with such remarkable accuracy that his Earth-centered model was still being used to predict the locations of the planets in the sky more than a thousand years after his death.

The ancient Indians of Asia had a different point of view, supposing that the Earth moves around the Sun, as did the Greek mathematician and astronomer Aristarchos, born on the island of Samos in 310 BC. Aristarchos moved the center of the Universe from the Earth to the Sun, and set the Earth in motion, supposing that the Earth and other planets travel in circular orbits around the stationary Sun, somewhat like heavenly moths around an outdoor lamp at night. He further stated that the fixed stars do not move, and that their apparent daily motion is due to the Earth's rotation on its axis.

As we now know, Aristachos was right. As the Earth rotates, the stars slide by, accounting for the wheeling night sky – so nicely described by the American poet Henry Wadsworth Longfellow:

> Silently, one by one,
> in the infinite meadows of Heaven.
> Blossomed the lovely stars,
> The forget-me-nots of the angels.[32]

And Aristarchos was also correct in supposing that the Earth revolves around the Sun, but his proposals made little impact on his contemporaries. It took another eighteen centuries before the Polish cleric and astronomer Mikolaj Kopernigk, better known as Nicolaus Copernicus, revived this heliocentric, or Sun-centered, model and the Italian astronomer Galileo Galilei provided observational evidence to support it. Galileo used the newly invented telescope to show that Venus goes through a complete sequence of Moon-like phases, from thin crescent to round disk. The full phase of Venus can only be seen from Earth when Venus is on the other side of the Sun, which meant that at least Venus has to go around the Sun.

In addition, Galileo demonstrated that Jupiter has four large moons, so not everything revolved around the Earth. And his telescope resolved the Milky Way into countless stars, which could not be seen with the unaided eye, suggesting that the stars have depth, with varying distances, and are not arranged along the smooth surface of a sphere.

But Galileo's support of Copernicus' theory, in which the Earth moves around the Sun, was opposed by theologians of the time. After trial by Inquisition in 1633, the Roman Catholic Church forced Galileo to recant his support of the Copernican system as "abjured, cursed and detested." He was banished to confinement at his house in Arcetri, in the hills surrounding Firenze, where he spent his last years. Legend has it that as Galileo rose from kneeling before his inquisitors, he murmured, "eppur, si muove" – "even so, it does move," but he would hardly have been foolish enough to risk even greater punishment. Not until 1992, more than 350 years after his trial, did Pope John Paul II in effect apologize for the harshness of Galileo's sentence.

Complete acceptance of the Sun-centered model required improved observations of the planets' movements, by Tycho Brahe; a crucial breakthrough by Galileo's contemporary, Johannes Kepler, who dispensed with circles once and for all, replacing them with ellipses; the discovery of the Sun's gravitational control of the planets by Isaac Newton, and eventual observation of the Earth's orbital motion around the Sun.

So the Earth is not the center, focus and fulcrum of all things. But we can understand people's reluctance to accept the idea. All of us are self-centered, and we usually imagine that we stand at an important, privileged location. By removing the center of the Universe from our home planet Earth, we seem to have become disoriented, losing our bearings and no longer at the center of everything.

A Delicate Balance – Gravity and Motion

The great English scientist Isaac Newton, born on Christmas day the year of Galileo's death, joined the Earth to the sky by showing that the same unchanging physical laws apply to the terrestrial and the celestial. He demonstrated that motions everywhere, whether up above in the heavens or down below on the ground, are described by the same concepts. So everything in the Cosmos doesn't just move, they move in predictable and verifiable ways.

According to tradition, Newton was sitting under an apple tree when an apple fell next to him on the grass. This reminded him that the power of gravity, whose pull influences the motion of falling bodies, seems undiminished even at the top of the highest mountains. He therefore argued that the Earth's gravitational force extends to the Moon, and showed that this force can keep the Moon revolving about our planet. It was as if the Moon is perpetually falling toward the Earth while always keeping the same mean distance from it. The Sun's gravity similarly deflects the moving planets into their curved paths, so they remain in perpetual motion around the Sun.

So Newton discovered the cosmic reach of gravity, which keeps our feet on the ground. Gravity has pinned us there, so we rotate with the spinning Earth and stay on it. The air and oceans are similarly held close to the planet by its gravitational pull.

The English genius was also a solitary loner, a self-isolated intellect, a bit obsessed, famously distracted, and frequently depressed. Newton didn't like interacting with people. He declined most invitations, avoided personal contact, never traveled abroad, and, they say, died a virgin at the age of eighty-five. He was also a Protestant theologian, a rebel against authority, and an enemy of the Pope. And Newton was enormously curious, once sticking a long needle along the edge of his eye just to see what would happen.

He spent much of his life immersed in experiments in alchemy and theological or mystical speculations, hoping to understand the origin of the elements and the eternal mysteries of health and mortality. The economist John Maynard Keynes described him as "the last of the magicians, the last of the Babylonians and Sumerians,"[33] trying to solve the riddles of the Universe by examining mystic clues left by God. And oh what a magician Newton was, conjuring up gravitation, the very force that rules the Universe.

It was his friend, the English astronomer Edmond Halley, who persuaded the secretive Newton to write his greatest work, the *Philosophiae naturalis principia mathematica,* or the *Mathematical Principles of Natural Philosophy,* commonly known as the *Principia.* It was presented to the Royal Society of London in 1686, which withdrew from publishing it owing to insufficient funds, so Halley, a wealthy man, paid for the publication the following year.

As Newton wrote in the *Principia,* "the Copernican system of the planets stands revealed as a vast machine working under mechanical laws here understood and explained for the first time." These were his laws of motion and the law of universal gravitation, achievements that resulted in Sir Newton becoming the first person in England to be knighted for his scientific work.

The basic ideas were that a body in motion stays that in motion unless acted upon by an outside force, and that gravity is the dominant force in the Cosmos. Although the weakest of the natural forces, gravitation exerts its influence to the greatest distance and has the capacity to act on all matter. That is why the force of gravity plays the central role in governing the orbits of the planets, as well as the motions of the stars and galaxies.

The enormous reach of gravity can be traced to two causes. The gravitational force decreases relatively slowly with distance, and this gives gravitation a much greater range than other natural forces, such as those that hold the nuclei of atoms together. And gravitation has no positive and negative charge, as electricity does, or opposite polarities as magnets do. This means that there is no gravitational repulsion between masses, no gravitational shielding, and every atom in the Universe feels the gravitational attraction of every other atom. In contrast, the attractive forces among unlike charges in an atom cancel each other, shielding it from the electrical forces of any other atom.

The gravitational force is mutual, so any two objects attract each other. The force of attraction increases with the mass of an object, and that mass possesses inertia, the tendency to resist any change in its motion.

Mass is incidentally an intrinsic aspect of an object, different from its weight that alters with distance from the main source of gravity. An astronaut weighs less when leaving the Earth, but retains the same mass.

The gravitational force between two bodies also weakens with increasing separation, and Newton used observations of planetary motions to determine just how the force decreases with increasing distance. Contrary to popular belief, Newton did not use his theory to show that planets move in elliptical orbits, but instead used Kepler's third law, connecting a planet's orbital period and distance, to show that the force of gravity must fall off as the inverse square of the distance.

Within the Solar System, the dominant gravitational force is that of the massive Sun, which is why we call it a Solar System. So it is the Sun's gravitational attraction that keeps the planets in their orbits, and holds the Solar System together. But why doesn't the immense solar gravity pull the entire planetary system into the Sun? Motion holds the planets up, opposing the relentless pull of the Sun's gravity and keeping the planets from falling into the Sun.

Each planet is moving at exactly the speed required to overcome the Sun's gravitational pull, keeping the planets in endless motion. If a planet moved any faster, it would leave the Solar System, and if it moved any slower it would be pulled into the Sun. This delicate balance between motion and gravity also explains why the Moon revolves around the Earth, and why Saturn's rings remain separate from the planet. The rings are made up of a vast number of particles, each one like a tiny moon revolving about Saturn.

Yet in Newton's time, no one knew how far the Earth is from the Sun, or how fast our planet moves through space, for the scale of our Solar System was unknown.

The Sun – How Far, How Big

Our planet glides through space at just the right distance from the Sun to keep most of our water liquid and for life to thrive. We sit in the comfort zone. Any closer and the oceans would boil away, as they did on Venus. Further out the Earth would be a frozen wasteland like Mars. But what is that critical distance from the Sun that permits life to thrive on our planet? Astronomers call it the astronomical unit, and they've spent centuries determining its precise value, inferring it from the distance between the Earth and a nearby planet, such as Venus or Mars.

A planet's distance can be determined by measuring the angular separation of the planet when observed simultaneously from two widely separated locations. This angle is known as parallax, from the Greek *parallaxis*, for the "value of an angle." If both the parallax and the separation between the two observers are known, then the distance of the planet can be determined by triangulation. It is based on the geometric fact that if you know the length of one side of a triangle and the angles of the two corners, all the other dimensions can be calculated.

The parallax technique of estimating a planet's distance is similar to the way your eyes infer how far away things are. To see the effect, hold a finger up in front of your nose, and look at your finger with one eye open and the other closed, and then with the open eye closed and the closed one open. Any background object near to one side of your finger seems to move to the other side, making a parallax shift. When this is repeated with your finger held farther away, the angular shift is smaller. In other words, the more distant an object, the smaller the parallax shift, and *vice versa*.

Venus moves closer to the Earth than any other planet, and the first reliable estimates for its distance were obtained in 1761 and 1769, when Venus moved directly in front of the Sun's disk. By timing the transits from stations at widely separated latitudes, astronomers estimated the planet's distance, and concluded that the astronomical unit is somewhere between 148 and 154 million kilometers with an uncertainty of about 10 percent.

Well-equipped expeditions were sent across the globe to observe these transits. The English astronomer Charles Mason and the surveyor Jeremiah Dixon sailed to Cape Town, South Africa, to observe the transit of 1761. The Royal Society of London sponsored Captain James Cook's expedition to observe the 1769 transit from Tahiti, which also enabled him to chart the coasts of New Zealand, Australia and New Guinea on the way home.

The most colorful transit story belongs to the French astronomer Guillaume Le Gentil. His trip to observe the 1761 transit from the French settlement at Pondicherry,

India was disrupted by a British blockade of the coastal fort. Le Gentil remained south of the equator, returning to Pondicherry for the 1769 transit, which could not be seen because of cloudy weather. To complete his string of bad luck, Le Gentil returned to France to find that he had been declared dead, with relatives dividing his belongings. But all things do not have to end badly; Le Gentil then married a wealthy heiress and had a satisfying career at the Observatoire de Paris.

A more precise measurement of the astronomical unit wasn't obtained for more than a century, in 1877 when Mars was closer to Earth than usual. David Gill, an unemployed Scottish astronomer with no university degree, traveled to the tiny island of Ascension near the equator, where he used the Earth's rotation to view Mars from different directions and infer its parallax. From this result, he determined a value of 149.8 million kilometers for the astronomical unit, with an uncertainty of about 200 thousand kilometers, and that's pretty close to the modern value of 149.6 million kilometers. Gill subsequently became Astronomer Royal at the Cape of Good Hope, and was eventually knighted, becoming Sir David Gill.

Although a more accurate value for the solar distance was sought using photographic observations of the asteroid Eros, when it passed near the Earth in 1930 – 31, significant improvements in precision only came in the late 1960s by bouncing pulsed radio waves off Venus and timing the echo. The round-trip travel time – about 276 seconds when Venus is closest to the Earth – was measured using accurate atomic clocks, and a precise distance to Venus was then obtained by multiplying half the round-trip time by the speed of light. This resulted in a precise determination of the astronomical unit at 149.597 870 million kilometers. At this distance it takes 499.005 seconds for light to travel from the Sun to the Earth, moving at the velocity of light 299,792 kilometers per second.

And once you have an accurate value for the Sun's distance, the Earth's mean orbital velocity can be determined, by dividing its orbital circumference by one year. That speed is tremendous, amounting to 29.8 kilometers per second, or 107 thousand kilometers per hour, about one twentieth the cruising speed of the supersonic jet aircraft Concorde.

How large and massive must the Sun be to supply its awesome energy, shining steadily and relentlessly over the eons? The size of the Sun can be determined from its angular extent and distance, obtaining a radius of 695.5 thousand kilometers or 109 times the radius of the Earth. And you can weigh the Sun from afar using the Earth's orbital period and distance, for it is the Sun's gravity that controls the Earth's orbital motion.

Substitution of the relevant numbers into Kepler's third law indicates that our home star has a mass of 1.989×10^{30} kilograms, or 333 thousand times the Earth's mass. As it turns out, the Sun contains 99.866 percent of all of the matter in the Solar System. The planets and everything else that orbits the Sun amount to only 0.134 percent of the total mass.

So the planets are just insignificant specks within the immensity of space, left over from the formation of the Sun and held captive by its massive gravity. This confirms the words of Marcus Aurelius written about 2,000 years ago: "The entire Earth is but a piece of dust blowing through the firmament, and the inhabited part of the Earth a small fraction thereof. In such a grand space, how many do you think will think of you?"[34]

This sense of the insignificance of planet Earth, and the people on it, was only further compounded when it was discovered that vast tracts of seemingly empty space separate the Sun from even the nearest stars. But this discovery followed the realization that the stars are moving through space.

3.3 THE DISTANT, MOVING STARS

Proper Motion

Despite nearly 2,000 years of stellar observations in antiquity, there was not a shred of evidence to contradict the ancient Greek belief that the stars are fixed, motionless with respect to each other and always displaying the same pattern in the sky.

Yet if the stars were not moving, their mutual gravitation would eventually pull them together into a single mass. Without motion, there would be nothing to keep the stars apart, and they should not be suspended in space.

Take globular star clusters as an example (Fig. 3.3). They consist of up to a million stars bound together into a spherical shape. And their individual stars must move here

FIG. 3.3   Globular star clusters A million stars are crowded together in globular star clusters, like the two shown here. Many are located in a great spherical halo that encloses our Milky Way Galaxy, gravitating about its massive center. A relatively few, like these, are concentrated toward the galactic nucleus. These globular clusters are designated NGC 6522 and NGC 6528, where NGC is an abbreviation for the *New General Catalog*, first compiled in 1888. (Courtesy of the Kitt Peak National Observatory and AURA.)

and there, like bees in a swarm, to counteract the gravitational pull of all the other stars and keep the cluster's shape. If the stars moved too slowly, they would be gravitationally pulled into each other, and the cluster would collapse like a failed soufflé. And if the stars moved at too fast a speed, they would disperse and the cluster could not hold itself together. So we know the globular cluster stars have to be moving at just the right speed to stay in the cluster and prevent its collapse, but the stars are crowded too close together to resolve individual ones and measure their motion.

Astronomers had to look at other closer stars to definitely show that stars move. Like a distant car on a highway, it's hard to judge a star's motion if it is headed straight toward or away from you. But if the star crosses at right angles to your line of sight, you ought to see it change its position in the sky. And assuming that all stars move through space at roughly the same speed, those closest to Earth should display the largest shift in position over a given length of time. A nearby bird flying overhead similarly travels rapidly across a great angle, while a high-altitude bird flying with the same speed barely creeps across the sky. That is how a duck hunter estimates the distance of a duck – by its angular speed.

Back in the third century BC, when no one had directly measured the motion of any star, the Greek astronomer Hipparchus compiled a catalogue of the positions and brightness of stars, which might be used to determine their motions in the future and to see if they changed in brightness or even faded away.

It took more than 1,800 years before anyone noticed that even one of the stars compiled by Hipparchus had moved. But when Edmond Halley compared his observations of stellar positions to those found in Ptolemy's reproduction of Hipparchus' catalogue, Halley concluded, in 1718, that at least three stars had changed position and moved.

The stellar motion that Halley detected is across the sky, transverse or perpendicular to the line of sight, and it is known as proper motion. This motion belongs to the star itself, in contrast to the improper motion caused by the Earth's movement in space. But proper motion isn't a velocity. It is the angular rate at which a star moves across the sky over the years or centuries, and it does not by itself determine the speed of motion. To convert a star's proper motion into a velocity, you have to know its distance, and no one knew the distance of any star other than the Sun in Halley's time.

As it turned out, the early measurements of proper motion were nevertheless a deciding factor in the first measurements of stellar distances.

Distance to the Nearest Stars

Although triangulation by parallax from widely separated places on Earth provided early estimates for the Sun's distance, no similar parallax could be detected for other stars. They are too far away for us to detect a shift in position from any two points on Earth. To triangulate their distances, astronomers needed a wider baseline – the Earth's annual orbit around the Sun.

When they observed stars at intervals separated by six months, astronomers could look at them from two points separated by a distance of twice that of the Earth from the Sun, or twice the astronomical unit. This might reveal a shift in the position of a star as seen from opposite sides of the Earth's orbit, a shift known as the annual parallax, and it could be used to calculate the distance of the star from the Earth.

A direct measurement of the true distance of any star, other than the Sun, wasn't easy, but a method for carrying out the delicate measurements had already been suggested by Galileo in the 17th century. It involves careful scrutiny of two stars that appear close together in the sky, one that is relatively nearby and the other much further away. The annual parallax of the nearest star can then be determined by comparing its position to that of the distant one for a year or more (Fig. 3.4). During the course of the year, the nearby star will seem to sway to and fro, in a sort of cosmic minuet that mirrors the Earth's orbital motion. The nearer the star, the larger the annual-parallax sway, but even the closest star was likely to exhibit only a modest shift, barely detectable at all.

If all the stars move with comparable speed, whether they are near or far away, then the star with the largest proper motion will be the closest one. And by the beginning of the 19th century it was known that one relatively undistinguished star has a proper motion that is significantly larger than almost every other star. This "Flying Star", known as 61 Cygni, became the first star, other than the Sun, whose distance was reliably determined.

With amazing fortitude, the German astronomer and mathematician Friedrich Wilhelm Bessel observed 61 Cygni and a fainter star near the same position in the sky, with the utmost care every clear night for more than a year, usually repeating his observations at least sixteen times a night. Then in December 1838 he announced that 61 Cygni weaved back and forth over a total angle of almost two-thirds of a second of arc, describing an annual ellipse that corresponds to a parallax of 0.31 seconds of arc, close to the modern value of 0.287.

The measurement indicated that 61 Cygni is about 700 thousand times further away than the Sun. Traveling at the speed of light, it takes 11.4 years to cross the distance from 61 Cygni to Earth. By way of comparison, it takes only 499 seconds for light to travel from the Sun to the Earth.

And the Sun remains a special star because of its proximity. Since the Sun is so much closer than the other stars, sunlight is millions of times more intense than all other starlight combined. That is why the day is bright and the night dark.

As the poet Francis Bourdillon expressed it in 1878:

The night has a thousand eyes,
And the day but one;
Yet the light of the bright world dies
With the dying Sun.

The mind has a thousand eyes.
And the heart but one;
Yet the light of a whole life dies
When love is done.[35]

The barrier to the stars had been broken. Astronomers finally knew with certainty the distance of at least one star, other than the Sun, and the sounding line to the stars had at last touched bottom. The result opened up the Cosmos to vast and extended horizons, placing even the closest stars at more than half a million times further away than the Sun, and demonstrating the enormous extent and roomy spaciousness of interstellar space. It also provided the first definite proof that the Earth really is in

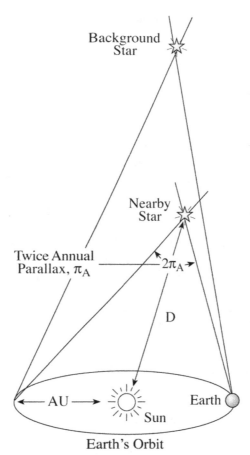

Earth's Orbit

FIG. 3.4 Annual parallax When a distant and nearby star are observed at six-month intervals, on opposite sides of the Earth's orbit around the Sun, astronomers measure the angular displacement between the two stars. It is twice the annual parallax, designated by π_A, which can be used to determine the distance, D, of the nearby star.

annual motion about the Sun, for you wouldn't have an annual parallax shift if the Earth weren't orbiting the Sun.

Within a century of Bessel's result, the annual parallax of about 2,000 nearby stars had been determined using long exposures on photographic plates, and the number tripled in succeeding decades. The closest star, known as Proxima Centauri, is 4.22 light-years away. And many of the brightest stars are hundreds of light-years away, so you can walk outside at night and see stars whose light was emitted before your parents were born.

Distant stars with parallaxes smaller than 0.05 seconds of arc cannot be measured with Earth-based telescopes because of atmospheric distortion that limits their angular resolution. However, instruments aboard the *HIPPARCOS* satellite, which orbited the Earth above its atmosphere in the 1990s, pinpointed the position of more than 100 thousand stars with an astonishing precision of 0.001 seconds of arc, determining the parallax and distance of many of them. That explains the spacecraft's name, which is an acronym for *HIgh Precision PARallax COllecting Satellite,* but the name also alludes to the ancient Greek astronomer Hipparchus, who recorded accurate star positions two millennium previously.

Once a star's distance is known, it can be combined with the star's observed luminosity to determine its intrinsic brightness. The technique is somewhat analogous to judging the distance of an oncoming car from the brightness of its headlights – the closer the car is the brighter its lights appear.

When the intrinsic brightness of stars was extrapolated from their apparent brightness and distance, astronomers were able to show that our eyes give us an incomplete view of the celestial realm. Just a few stars that appear bright to the eye are relatively nearby – Sirius is an example. The majority of the apparently brightest stars are far-flung, bloated giant stars that pour fourth thousands of times more energy every second than the Sun, making them stand out at night. When telescopes were used to make a complete stellar census, it was found that the most common stars are not bright. They are so dim that they cannot be seen with the unaided eye.

3.4 SIZE AND SHAPE OF THE MILKY WAY

A Fathomless Disk of Stars

All the stars that you can see on a clear moonless night belong to the hazy, luminous band of starlight known as the Milky Way (Fig. 3.5). According to ancient Greek myth, the infant Heracles, the son of Zeus, bit the breast of his mother, the goddess Hera, and spilled her milk into the sky. The Romans called the spilt milk the *Via Lactea,* or the "Milky Way." It is also called our Galaxy, with a capital G, derived from the Greek word *galakt-* for "milk."

The Milky Way has long been thought to be a path to the heavens. In his *Metamorphoses,* written in the first century, the Roman poet Publius Ovidius Naso, or Ovid for short, described it as a passageway to the homes of the gods, or in modern terms, an escalator into the sky or a celestial highway. Sixteen centuries later, the English poet John Milton described the Milky Way as "a broad and ample road, whose dust is gold" and is "powdered with stars."[36]

The brightest stars, which are most easily detected, are too close for perspective. Our eyes, for example, detect almost nothing within or behind the constellations that would give them a three-dimensional configuration. And astronomers cannot look at the Cosmos from outside, because our planet and the Sun are immersed within it.

We can nevertheless speculate about the structure of our stellar system, and an early, perceptive interpretation was provided in 1758 by the astronomer Thomas Wright, born near Durham in the north of England. Like the ancient Greeks, Wright

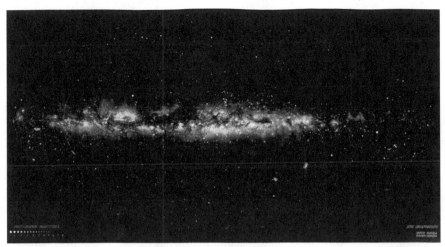

FIG. 3.5   The Milky Way A panoramic telescopic view of the Milky Way, the luminous concentration of bright stars and dark intervening dust clouds that extends in a band across the celestial sphere. The Milky Way defines the plane and main disk of our Galaxy. We live in this disk, and look out through it. Our view is eventually blocked by the build-up of interstellar dust, and the light from more distant regions of the disk cannot get through. The center of the Galaxy is located at the center of the picture, in the direction of the constellation Sagittarius. Although the disk appears wider in that direction, the galactic center is not visible through the dust. The Large and Small Magellanic Clouds, companion galaxies to the Milky Way, can be seen as bright swirls of light below the plane to the right of center. (This map of the Milky Way was hand-drawn from many photographs by Martin and Tatjana Keskula under the direction of Knut Lundmark; courtesy of the Lund Observatory, Sweden.)

supposed that the Universe is spherical, but unlike them, he placed the Sun with all the other stars in a thin spherical shell or bubble. Its radius was imagined to be so great that the part we are immersed in, and which is near enough to see, is just a small segment of the shell, so tiny that it looks like a flat collection of stars, the Milky Way (Fig. 3.6).

In an effort to reconcile his telescopic observations with his religious beliefs, Wright assumed that the Divine Presence occupied the center of the stellar sphere, or to be exact, our God was centered in our stellar sphere, since there might be other spheres with other centers, the abodes of different gods. Darkness resided outside our shell, perhaps together with some nebulae.

Although Wright's speculations about spheres and their divinities are not accepted today, his model of our local stellar system is. We are located within the disk and plane of the Milky Way, viewing it edgewise. When gazing directly into this stellar layer, the light of numerous stars illuminates the Milky Way, and outside this direction relatively few stars are found, set against empty and dark space. It is similar to living in a city, noticing buildings all around you, but none when you look up into the sky.

Just a little more than two decades after Wright's proposals, the German-born English astronomer William Herschel discovered Uranus, receiving a lifelong pension from King George III. With the king's support, Herschel constructed the biggest tele-

Sphere of Stars   Milky Way

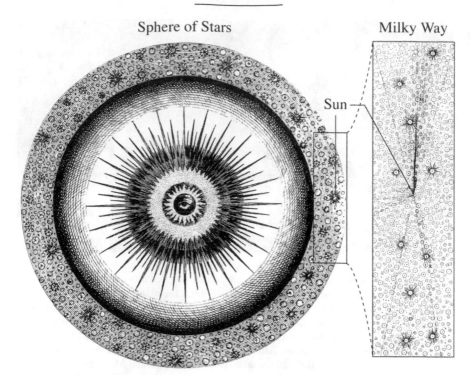

Sun

FIG. 3.6   Thomas Wright's model of the Milky Way According to Thomas Wright of
Durham (1711–1786), the Milky Way is composed of a large number of stars arranged
in a slab-like layer, with the Sun placed at the center of the slab. A cross section of this
stellar layer is shown here. It is a small segment of a much larger spherical shell of stars,
with the Divine Presence signified by an eye located at the center. (Adapted from Thomas
Wright, *An Original Theory or New Hypothesis of the Universe*, 1750, Reproduced by Science History Publications, New York 1971, page 139 plate XXV.)

scopes of the time, with the largest mirrors and greatest light-gathering power, supposing that they would enable him to plumb the very depths of the Milky Way.

His efforts culminated with his 1.2-meter (48-inch) mirror, weighing nearly a ton
(1,000 kilograms). It was housed within a 40-foot-long metal tube within a wooden
framework that could roll around a low, circular wall of masonry. After orientation
within that circle, the tube was raised to an appropriate height with the aid of ropes and
pulleys, like hoisting the sails of a boat or raising a huge cannon up to shoot the stars.
Then William, working with his sister Caroline, would just let the sky drift by as the
Earth rotated and the night passed away, counting the number of stars visible in each
place they were looking. They thereby hoped to detect the faintest and most distant
stars, determining how far the stars extend into space in different directions.

Herschel's 40-foot reflector, as it was called, was the largest telescope of its time,
and remained so for half a century. It soon gained a reputation as the "eighth wonder
of the world." Since it was built with the king's money, he was present at the dedication,
telling his guest, the archbishop of Canterbury, that the telescope would show him the

way to heaven. And Franz Joseph Haydn's visit to the colossal telescope was apparently the direct inspiration for "The Heavens Are Telling" passage in his great oratorio *The Creation*.

But the giant telescope still wasn't big enough to fathom the "profundity" or depth of the Milky Way. It just brought more faint stars into view, and pushed the edge of the known Universe further into space. And astronomers never have succeeded in deciphering the true shape and size of the Milky Way by observing its stars, even when looking much further into it using larger telescopes and photographic techniques that permitted long exposures. That is because the most powerful telescopes can only discern the visible parts of our stellar system, and not its most distant, invisible parts that lie behind an opaque veil of interstellar material.

Looking deep into the Milky Way is something like viewing distant objects on a foggy day. At a certain distance, the total amount of fog you are looking through mounts up to make an impenetrable barrier. The fog then becomes so thick and dense that it blocks the light of distant stars. We can only see that far, and more distant objects are hidden from view. New perspectives were required to look around, and eventually through, this barrier to the heavens.

The Sun is Off-Center

The true enormity of our stellar system was discovered by using Cepheid variable stars to gauge distances. The Cepheids, named after the first of the type to be discovered, Delta Cephei, are yellow supergiant stars that periodically brighten and dim with a period of days, over and over again like stellar fireflies.

At the end of the nineteenth century, Harvard College established an observatory at Arequipa, Peru, in order to extend its photographic surveys to the entire sky. Between 1893 and 1906 the two Magellanic clouds were systematically recorded with the 0.60-meter (24-inch) refractor at Arequipa, and from these plates Henrietta Leavitt, a researcher in Cambridge, Massachusetts, found an extraordinary total of 1,777 variable stars. By 1908 she had derived periods for a few of these Cepheid variables, and she reported that the brighter stars tended to have the longer cycles of variation. Because the extent of the Magellanic clouds is small compared to their distance, the relation of period to apparent brightness implied also a real connection with absolute brightness or luminosity. And in 1912 Leavitt had obtained apparent brightness and period data for 25 variable stars in the Small Magellanic Cloud, thereby establishing the important period-luminosity relation for the Cepheid variables.

Once this period-luminosity relation is suitably calibrated by an independent measurement of the distance of just one variable star, observation of the period of any other Cepheid variable leads to a determination of its intrinsic luminosity. When combining this with the observed brightness, the distance of the star can be calculated. And because the Cepheid variable stars are very bright, averaging about 10 thousand times the brightness of the Sun, they are conspicuous and can be seen to exceptional distances, well beyond the reach of annual parallax methods of determining stellar distance.

It was the American astronomer Harlow Shapley who observed Cepheid variables in the globular star clusters, located outside the plane of the Milky Way, discovering in 1918 that the Sun if far removed from the center of the stellar system. By investigating these star clusters, he was looking up, out and beyond the known stellar Universe. Using the 1.5-meter (60-inch) telescope on Mount Wilson, California, the largest of its time, Shapley could easily identify the varying stars, which stood out amongst the hundreds of thousands of other stars in each cluster like a searchlight or airport beacon within the lights of a city. Then, using the period-luminosity law of Cepheid variation, Shapley established the enormous distances of the globular star clusters, and also showed that they are distributed within a roughly spherical system, which is centered far from the Sun in the direction of the constellation Sagittarius (Fig. 3.7).

Shapley rolled back the boundary of the known Universe and placed the Sun in a remote corner of it. Modern determinations indicate that the center of our stellar system is at a distance of 27,700 light-years from the Sun, and the radius of the highly flattened disk of the Milky Way is about twice that amount. So the Milky Way is a truly big and lonely place, billions of times larger than the distance between the Earth and the Sun and thousands of times bigger than the distance to the nearest star other than the Sun.

What accounts for the shape of the Milky Way? In 1781 the German philosopher Immanuel Kant supposed that it was flattened by its rotation, and although they didn't at first realize it, astronomers could detect that rotation by observing the motions of nearby stars. They move across the sky in two intermingled streams, that apparently flow in opposite directions.

The two star streams are explained if all the stars in the Milky Way are whirling about the remote, massive galactic center. The enormous mass at this central hub steers the stars into circular orbits with an orbital speed that decreases with increasing distance from the center. Stars in orbits inside that of the Sun travel faster and so forge ahead of us, while the stars moving in orbits outside the Sun's are falling behind. When viewed from the Earth, nearby stars that are a little closer than the Sun to the center therefore seem to move in one direction, while those a bit farther away appear to move in the opposite direction.

FIG. 3.7   Edge-on view of our Galaxy The globular star clusters are distributed in a roughly spherical system whose center coincides with the core of our Galaxy. The Sun is located in the galactic disk, about 27,700 light-years away from the center of our Galaxy. The galactic disk and central bulge are shown edge-on as a negative print of an infrared image taken from the

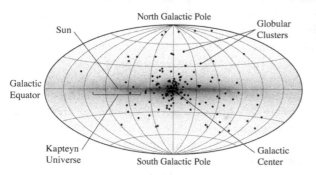

InfraRed Astronomical Satellite. The infrared observations can penetrate the obscuring veil of interstellar dust that hides the distant Milky Way from observation at visible wavelengths. It is this dust that limited astronomer's view of the surrounding stars to a much smaller Kapteyn Universe, centered on the Sun.

We can weigh the massive center using the rotation speed and distance of the Sun about it. That central mass is equivalent to about 100 billion times the mass of the Sun. And this means that the Sun is one of about one hundred billion stars in our Milky Way.

Whirling Coils of the Milky Way

The stars do not reside in a uniform whirling disk, like a spinning plate or Frisbee. They are instead concentrated into four arms that coil out from the galactic center, giving our stellar system a spiral shape. These arms were first delineated by relatively young, very bright and massive stars, which light up the nearby arms like city streetlights seen from an airplane.

Because the Sun is embedded in one of the arms (Fig. 3.8), astronomers have to look through that arm to see the rest of the Milky Way, and this obscures their distant vision, hiding most of our stellar system from view in visible light. But radio waves pass unimpeded through the obscuring material, permitting the detection of most of the Milky Way. By observing the radio emission of interstellar hydrogen atoms, radio astronomers have delineated extensive, arm-like concentrations that extend out from the short segments defined by young massive stars in the vicinity of the Sun, describing the whirling spiral coils of the Milky Way (Fig. 3.9).

FIG. 3.8 Our place in space In this painting, the Milky Way and its massive center are depicted from slightly above the plane of our Galaxy. Bright stars are concentrated within spiral arms that wind out from the central bulge. Since the Sun lies within one of the spiral arms, we look into them edgewise and see the nearby ones as the luminous band of the Milky Way. (Courtesy of Jon Lomberg.)

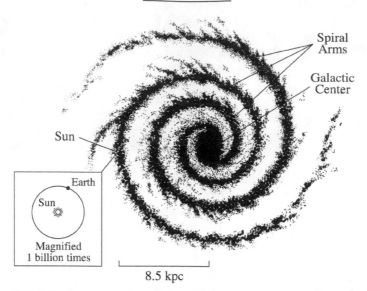

FIG. 3.9   **Structure of our Galaxy** This drawing depicts our Galaxy as viewed from above the Milky Way or galactic plane. The stars and interstellar material are concentrated within spiral arms. The Sun lies within one of these spiral arms at a distance of 27,700 light-years from the center of our Galaxy, designated here as 8,500 parsecs or 8.5 kpc. This distance is 1.75 billion times the distance between the Earth and the Sun.

But astronomers had already realized that other cosmic objects, known as spiral nebulae, have a spiral shape. These spirals were eventually shown to be remote galaxies, lying far beyond the edges of our Milky Way.

3.5 GALAXIES AND THE EXPANDING UNIVERSE

Spiral Nebulae – Discovery and Fleeing Motion

When William Parsons, third Earl of Rosse built the largest telescope of the time at Birr Castle in Northern Ireland (Fig. 3.10), he thought that it might resolve an ongoing controversy about the diffuse patches in the night sky known as *nebulae*, the Latin word for "clouds." The famous English astronomer, William Herschel, had attributed their cloud-like, nebulous forms to some sort of dispersed, shining, fluid-like material, but other astronomers thought the nebulae were composed of stars that are too far away to be detected and resolved by small telescopes.

But the unsurpassed light-gathering power of Lord Rosse's 1.8-meter (72-inch) metallic mirror only compounded the problem of the nebulae. It revealed a whole new class of the objects, called the spiral nebulae. Rosse worked from a catalogue of more than 100 prominent nebulae assembled by the French comet hunter Charles Messier to avoid misinterpreting them as comets. And within weeks of first pointing his telescope at the sky in 1844, Rosse found that the fifty-first nebula in Messier's list, designated

FIG. 3.10 Leviathan of Parsonstown William Parsons (1800–1867), the third Earl of Rosse, surveyed the heavens with this telescope, from the grounds of Birr Castle, his family's ancestral estate in Northern Ireland. The four-ton, 1.8-meter (72-inch) speculum mirror, made of an alloy of copper and tin, was the largest in the world from 1845 to 1977, when it was surpassed by the 2.5-meter (100-inch) Hooker telescope at Mount Wilson Observatory near Pasadena, California. Lord Rosse's telescope, also known as the "Leviathan of Parsonstown", is remembered for its ability to resolve some nebulae into a spiral shape (Fig. 3.11). The mirror was mounted at the bottom of a 15-meter (58-foot) wooden tube slung between two massive stonewalls. Assistants would hoist the tube to a particular angle using an unwieldy system of cables, and this permitted observations of a strip of sky as it crossed the meridian. (Courtesy of the Trustees of the Science Museum, London.).

M51, exhibits a spiral pattern, which he attributed to its rotation (Fig. 3.11). This spiral nebula is now commonly known as "The Whirlpool Nebula", suggesting a whirling vortex of light.

The devastating potato famine in Ireland interrupted Rosse's astronomical work, which he abandoned to devote himself to his sick and dying tenants. But he returned to his telescope three years later, and eventually found fourteen nebulae that exhibit a spiral pattern.

Rosse's drawing of M51 was widely reproduced, appearing in *Les Etoiles,* a popular book published in 1882 by the French astronomer Camille Flammarion. And it is possible that the Dutch artist Vincent Van Gogh saw the shape there, reproducing it in his second portrayal of *The Starry Night* executed from Saint-Rémy in 1889 (Fig. 3.12).

FIG. 3.11   Spiral shape of Messier 51 Lord Rosse discovered the spiral structure of
Messier 51, abbreviated M51, using his 1.8-meter (72-inch) telescope (Fig. 3.10) in the
spring of 1845, and he subsequently found thirteen other nebulae with a spiral shape. In
his description of this drawing of M51, published in 1850, Rosse attributed the spiral pattern to rotation of the nebula. Camille Flammarion (1842–1925) included this sketch in
his popular books about astronomy, leading to a growing awareness of spiral nebulae, and
it might have inspired the swirls of starlight found in Vincent Van Gogh's (1853 – 1890)
painting of the *Starry Night* (Fig. 3.12). We now know that M51, also designated NGC
5194 and called the Whirlpool Galaxy, is a magnificent, rotating spiral galaxy located 35
million light-years away (Fig. 3.13), with a small, irregular companion NGC 5195 separated from the center of M51 by about 10 million light-years. [Reproduced from The Earl
of Rosse, "Observations of the Nebulae", *Philosophical Transactions of the Royal Society*,
pages 110 – 124, plate 35 (1850).]

Van Gogh's rendition of the active, swirling form of a spiral nebula plays a dominant
role in the painting, and it bears a striking resemblance to Flammarion's description
of M51, which "flows like a celestial river" and "seems to tear into shreds like a fleece
combed by the winds of the sky."[37]

Near the end of the 19th century, James E. Keeler began systematic photography of
the nebulae with the 0.91-meter (36-inch) Crossley reflector at the Lick Observatory in
California, discovering thousands of spiral nebulae. He estimated that 120,000 spirals
could be seen with the telescope, always avoiding the plane of the Milky Way.

FIG. 3.12 **The starry night** The intense color and restless feeling in this visionary canvas captures the spiritual power of an incandescent night sky, which is filled with sweeping, rolling shapes. It was painted by the Dutch artist Vincent Van Gogh (1853–1890) at Saint-Rémy, France in mid-June 1889. Enormous stars and the vast and graceful swirls of a spiral nebula loom over a diminutive church spire, while the top of a dagger-like cypress tree links the Earth and humanity to the stars. Much of the painting is a composite portrayal of the landscape and sky Van Gogh would have seen from his window and from the neighborhood of the asylum where he was staying in June of that year, including the lunar crescent and the planet Venus, the bright object near the horizon. The undulating band of light is probably the Milky Way, and the stars above it may be the shaft and one arm of the Northern Cross, which he would have seen in the east during the evening. The shape of the swirling pattern resembles Lord Rosse's drawing of the spiral nebula M51 (Fig. 3.11), but reversed as in a mirror so the nebula in the painting seems to whirl clockwise. Oil on canvas, 0.737 × 0.921 m. (Courtesy of the Museum of Modern Art, New York, acquired through the Lillie P. Bliss Bequest.)

Only one of the nebulae could be seen with the unaided eye, the Andromeda Nebula, or M31, a faint, fuzzy glow in the constellation Andromeda. Thus an entirely new Universe of more than a hundred thousand spiral nebulae was revealed, almost entirely invisible to the unaided eye. And we still marvel at their beauty, revealed by the most sophisticated telescopes of our time (Fig. 3.13).

At the time of their discovery in enormous numbers, most astronomers thought that the spiral nebulae were nascent planetary systems. The bright center was supposed to be a newborn star, and the spiral arms surrounding it were developing planets, whirling and rotating around the central star just as the Earth revolves around the Sun.

FIG. 3.13 The Whirlpool Galaxy The *Hubble Space Telescope*
reveals the details of the Whirlpool Galaxy, also known as M51 and
NGC 5194, a spiral galaxy located about 35 million light-years away,
near the Big Dipper in the constellation Canes Venaciti. This image
should be compared with Lord Rosse's drawing of M51 made more
than 150 years before (Fig. 3.11). A smaller, nearby companion galaxy,
NGC 5195, is just off the corner of this image, but it is seen in Fig.
3.11. The companion's gravitational pull is apparently triggering the
formation of massive, luminous stars from hydrogen gas in the spiral
arms of M51. Delicate structure is seen in the cold dust clouds, with
spurs branching out almost perpendicular to the arms. (Courtesy of the
Hubble Heritage Team, NASA, STScI, and AURA.)

The wealthy Bostonian Percival Lowell had built an observatory in Tucson, Arizona, primarily to detect canals supposedly built by parched, industrious Martians. He also believed that the spiral nebulae resemble our Solar System in its early formative stages, and therefore instructed his staff astronomer, Vesto M. Slipher, to measure their rotations, hoping to gather insight about our own planetary system.

As often happens in astronomy, Slipher's measurements resulted in an entirely unexpected discovery. When using the 0.6-meter (24-inch) refractor at the Lowell Observatory to record the spectra of bright spiral nebulae, he found that they are almost unanimously moving away from us at high velocities. They were also rotating, and a few were approaching, but at modest speeds in comparison to the outward motion of most spirals. The Andromeda Nebula was, for example, moving toward the Earth at an apparent velocity of 300 kilometers per second. But all the other bright spirals were moving in the opposite direction, usually with much larger velocities of up to 1,100 kilometers per second, much faster than any star in the Milky Way.

By 1917, Slipher had accumulated spectra of 25 spiral nebulae, using the Doppler effect to measure their radial velocities, and showed that none of them are at rest. All but three were rushing away from us, and each other, dispersing, moving apart and occupying an ever-increasing volume, like a puff of smoke dispersing into the air.

According to Slipher, the observed motions of the spirals indicated "a general fleeing from us or the Milky Way." And it was certainly hard to believe that objects with such enormous speeds could long remain a part of the Milky Way system. The combined gravitational pull of the entire 100 billion stars in our Galaxy is not enough to retain any spiral nebula moving at speeds in excess of 1,000 kilometers per second.

Today, when larger telescopes and electronic detectors are readily available, it is difficult to comprehend how much time Slipher required in order to record the spectra of even the most luminous spiral nebulae. The spirals were faint, and the light that reached Earth was further reduced in intensity to near-invisibility when dispersed to obtain their spectra. Heroic exposure times of 20, 40 and even 80 hours were needed to obtain the elusive spectra, and to infer radial velocities from the Doppler shifts of the spectral lines recorded in them. Subsequently, bigger telescopes collected more light, thereby reducing the necessary observing time for the brighter spirals. The larger telescopes could also see further into space, where the fainter spirals are located, and they could resolve individual components of the nearby spirals, which eventually led to measurements of their distances.

Bigger Telescopes

No one realized the possibilities of large telescopes more than George Ellery Hale, probably the greatest scientific entrepreneur of the 20th century. While a young Associate Professor at the University of Chicago, Hale convinced the wealthy Charles T. Yerkes that he should endow the Yerkes Observatory, funding the construction of the world's largest refractor there, with a lens 1-meter (40-inches) in diameter. Then early in the 1900s Hale moved to Pasadena, California, near Los Angeles, founding a solar observatory on nearby Mount Wilson, which rose above the dense smog below. A few years later, Hale inaugurated a new telescope on Mount Wilson to look at the entire Cosmos, using a 1.5-meter (60-inch) mirror provided by Hale's father.

Hale then persuaded John D. Hooker, a Los Angeles businessman, to pay for a 2.5-meter (100-inch) mirror, weighing 4.5 tons, and installed it on Mount Wilson with funding by the Carnegie Institution of Washington, D.C. The Carnegie Institution is a private foundation supported largely from endowments by Andrew Carnegie, the former owner of the Carnegie Steel Company, once the richest man in the world, and the philanthropist who built more than 2,500 libraries in the English-speaking world. The Carnegie Institution also funded the Mount Wilson Observatory, a sort of gentleman's club that ran the 100-inch Hooker telescope for three decades, when it was used to discover the galaxies and the expanding Universe.

And Hale, the indefatigable fund-raiser, didn't stop with the 100-inch, writing a stirring call for a yet larger telescope, published in 1928 in *Harper's Magazine*. It included:

> Like buried treasures, the outposts of the Universe have beckoned to the adventurous from immemorial times. Princes and potentates, political or industrial, equally with men of science, have felt the lure of the uncharted seas of space, and through their provision of instrumental means the sphere of exploration has rapidly widened. . . . Each expedition into remoter space has made new discoveries and brought back permanent additions to our knowledge of the heavens.[38]

Hale asked the editor of *Harper's* to send a copy of his article to the Rockefeller Foundation, where it had a catalytic effect in bringing forth six million dollars for building a 5.0-meter (200-inch) telescope, finished in 1948 on Palomar Mountain in California. This instrument reigned as the unrivaled leader in astronomy at visible wavelengths for almost half a century, peering deeper into the Universe than ever before and playing an important role in contemporary discoveries of the violent Universe.

Hale also brought the California Institute of Technology, abbreviated Caltech, into existence in 1920, and the Rockefeller's preferred to give their money to the university. So Caltech owned the 200-inch telescope, and the Carnegie Institution owned the 100-inch telescope. So the two institutions ran both telescopes together as the Mount Wilson and Palomar Observatories, until the 1960s when they were renamed the Hale Observatories. But it was the 100-inch Hooker telescope that resulted in the most profound discoveries, at about the same time that Hale was dreaming of bigger telescopes to come.

Discovery of the Galaxies

Edwin P. Hubble, a brilliant, colorful and polished young man, had already turned down the chance to become a professional boxer, served as one of the first Rhodes scholars at Oxford, obtained a doctorate at the University of Chicago, and served briefly in the Army, when in 1919, at the age of thirty, he moved to California and took an astronomy position at the Mount Wilson Observatory. As one of the first astronomers to use the new 100-inch Hooker telescope, Hubble eventually settled an ongoing controversy over the nature of spiral nebulae.

The issue was presented during the now-famous Shapley-Curtis debate over "The Scale of the Universe" during a meeting of the National Academy of Sciences held on 26 April 1920 at the Smithsonian Institution in Washington, D.C.. Harlow Shapley defended his novel conception of a much larger Milky Way than had previously been supposed, with a distant center and the Sun at its periphery, but supposed that the spiral nebulae are embedded in the cozy confines of the Milky Way. In contrast, Heber D. Curtis, of the Lick Observatory, attempted to defend a smaller Sun-centered stellar system, but provided cogent arguments that the spiral nebulae are distant stellar systems located far beyond the Milky Way. Shapley was right about the shape and size of the Milky Way, but Curtis was correct in supposing that the spiral nebulae are distant "island Universes" composed of numerous stars.

The argument over the extragalactic nature of the spiral nebulae was finally and definitely resolved when Hubble used the Hooker telescope to photograph the spiral nebula Andromeda, or M31, night after night, comparing hundreds of photographs to find Cepheid variable stars whose brightness waxed and waned like clockwork with a lengthy period of several days. On New Year's Day, 1925, Hubble's results were read *in absentia* at a meeting of the American Astronomical Society in Washington, D.C., causing an overwhelming sensation. The landmark paper, entitled "Cepheids in Spiral Nebulae", combined the known period-luminosity relation of Cepheids with observations of these variable stars in M31 and M33, another spiral nebula, to derive a distance of 930 thousand light-years for the two spiral nebulae. They had to be remote objects ablaze with stars, galaxies in their own right and separated from the Milky Way by wide gulfs of apparently empty space.

So the Milky Way had become just one Galaxy, with a capital G to show that it is ours and therefore special. And the spiral nebulae became known as spiral galaxies, the designation nebulae now being reserved for cloudy, gaseous material enveloping bright stars.

The Milky Way no longer contained everything there is, and the Universe was opened wide up. It was no longer limited to the things our unaided eyes can focus on, and our stellar system became just one of myriads of galaxies that stretch as far as the largest telescopes can see – and even beyond.

It is as if the galaxies had broken free of any local constraints, reminding us of the poem written more than eight centuries ago by the Japanese Zen Buddhist Eihei Dogen:

> Four and fifty years
> I've hung the sky with stars.
> Now I leap through –
> What shattering![39]

Like the poet, Hubble broke through the stars, and moved the outer boundaries of the Universe far out into space, enlarging our horizons and perhaps removing our feelings of constraint.

When looking up at the night sky, you will only see stars, and the black spaces between them. The galaxies are out there, but you cannot see them without a telescope. They are so far away that the light from the most distant ones was generated before the Earth was born. And even more fantastic, they are all in flight, rushing away from us at speeds that increase with their distance.

The Universe is Expanding

After using Cepheid variable stars to infer the distances of just a few nearby spiral nebulae, Hubble estimated the distances of several other galaxies by assuming that their brightest stars are intrinsically as bright as the most luminous stars in the Milky Way. As stellar radiation propagates out into space, its luminosity falls off with the square of the distance, which can be determined from the ratio of the intrinsic and observed brightness of a star.

By 1929, Hubble had approximate distance measurements for just 24 spirals, and he compared these distances to their redshifts, or radial velocities, mainly provided by

Vesto Slipher with a few, highly crucial, larger redshifts measured by Hubble's colleague Milton Humason. The comparison indicated that the more distant a galaxy is, the faster it is rushing away from us.

In his publication of these results, entitled "A Relation between Distance and Radial Velocity Among Extra-Galactic Nebulae," Hubble drew a straight line through a plot of the observed data, but there was a wide dispersion between the plotted points and only a mild tendency for velocity to increase with distance (Fig. 3.14). Nevertheless, his conclusion was subsequently confirmed by more comprehensive observations of a much greater number of galaxies.

The connection between velocity and distance is now known as Hubble's law, and the ratio of the velocity of recession of any galaxy and its distance from us is now called Hubble's constant, a fundamental measure of the Universe.

The units of Hubble's constant are given in kilometers per second per Megaparsec, and Hubble's initial estimate was pegged at 530 in these units. The radial velocity is in units of kilometers per second, and the distance is in Megaparsecs, abbreviated Mpc, where one Mpc is equivalent to 3.26 million light-years. Galaxies are typically separated by a few Mpc, or about 10 million light-years, which is about 100 galaxy diameters. So the Universe is largely empty space as far as galaxies are concerned.

In just two years, Hubble extended the velocity-distance relation to substantially greater distances, with the help of Humason who made the velocity measurements. Humason was a tobacco-chewing gambler and reputed "ladies-man" whose formal education ended in the eighth grade, beginning his career on Mount Wilson as a mule

FIG. 3.14   Discovery diagram of the expanding Universe A plot of the distance of extra-galactic nebulae, or galaxies, versus the radial velocity at which each galaxy is receding from Earth, published by Edwin Hubble in 1929. The approximate linear relationship between the

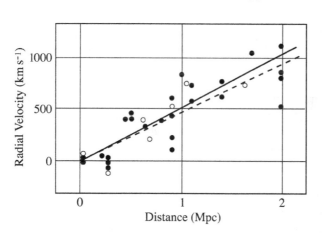

distance and radial velocity indicates that the Universe is expanding. Here the velocity is in units of kilometers per second, abbreviated km s^{-1}, and the distance is in units of millions of parsecs, or Mpc, where 1 Mpc is equivalent to 3.26 million light-years. Hubble underestimated the distances of the spiral nebulae, so the distance scale for modern versions of this diagram is about seven times larger (Fig. 3.15). The filled circles and solid line represent the solution for individual nebulae; the open circles and dashed line are for groups of them, the cross represents the mean distance of 22 nebulae whose distances could not be determined individually. [Adapted from Edwin P. Hubble, "A Relation between Distance and Radial Velocity among Extra-Galactic Nebulae", *Proceedings of the National Academy of Sciences* **15,** 168 – 173 (1929).]

driver and observatory janitor. But after filling in as a substitute night assistant, he became an astronomer more skilled at observing than Hubble when it came to obtaining the spectra needed to infer a radial velocity from the Doppler effect. Even with the 100-inch telescope, the velocity measurements of these remote galaxies required photographic exposures up to fifty hours long.

Because of the impossibility of measuring distances to such faint objects, Hubble and Humason simply assumed that all galaxies have the same intrinsic luminosity, and inferred a velocity-distance relation by comparing the observed velocities of the galaxies to their apparent brightness. They thus reformulated Hubble's law and showed that a linear relation is valid, within the observational uncertainties, out to radial velocities of nearly 20,000 kilometers per second and distances as large as 100 million light-years.

Humason then teamed up with Nicholas Mayall at the Lick Observatory in an ambitious 25-year project of painstakingly measuring the Doppler-effect redshifts of nearly a thousand galaxies visible from the northern hemisphere. Mayall used the venerable 36-inch (0.9-meter) Crossley reflector on Mount Hamilton to observe the brighter galaxies, while Humason observed the fainter ones using the 100-inch Hooker telescope on Mount Wilson.

Although Hubble had initiated the project, he died before the work was finished, and the analysis was left in the hands of his young protégé Allan Sandage, who obtained a value for the Hubble constant of 180 kilometers per second per Megaparsec, or about one third the value previously found by Hubble, whose distance scale was in error. His mistake was not discovered until the early 1950s when red-sensitive photographic plates, developed for military reconnaissance in World War II, became routinely available. Sandage then used them to discover that the brightest stars, which Hubble used to infer distances, are much more luminous emission nebulae.

The 1956 publication by Humason, Mayall and Sandage included a redshift-apparent brightness diagram of 474 galaxies, or extra-galactic nebulae as they preferred to call them. Although the plot had a large scatter, they could draw a straight line through it, out to a radial velocity of 100,000 kilometers per second, or one-third the speed of light. So Hubble's law connecting radial velocity and distance still held in every direction as far as one could see. This meant that the entire Universe is expanding, swiftly and evenly in all directions, with the fastest-moving parts having traversed the greatest distance.

Hubble never said much about the expansion of the Universe, but the discovery made him famous. He appeared on the cover of *Time* magazine, on the 9 February 1948 issue, and was photographed with Hollywood stars. Scientific recognition also followed, but never the Nobel Prize. Hubble had not only extended the observable Universe far beyond our stellar system, the Milky Way, but he had also shown that the entire Universe is expanding. These accomplishments were certainly worthy of a Nobel Prize, and Hubble was nominated for it. At Hubble's time, however, the Nobel Prize in Physics was not extended to astronomy, and there is no Nobel Prize in Astronomy. The international awards have been given yearly since 1901 by the Nobel Foundation, which was endowed in 1895 in the will of the Swedish industrial chemist and philanthropist Alfred Nobel, the inventor of dynamite. Because Nobel's wife had run off with one of the leading astronomers and mathematicians of the time, his bequest specified that neither astronomy nor pure mathematics could be considered.[40] Nobel did not want to provide a way of recognizing the man, who might have been a candidate for one of the first prizes.

As Hubble realized, astronomers only see as far as their telescopes permit, eventually reaching a limit, a dim boundary to the observable Universe where they "measure the shadows." And even now, there is no telescope powerful enough to detect the edge where the galaxies might end.

Hubble relied solely on the power of observation, preferring to avoid what he called "the dreamy realms of speculation,"[41] most likely referring to theoretical physicists. Although an expanding Universe was a possible consequence of Albert Einstein's *General Theory of Relativity*, Hubble thought that such models were a forced interpretation of the observational results. Even as late as 1953, shortly before his death by heart attack, Hubble wrote about his law of the redshifts in the *Monthly Notices of the Royal Astronomical Society*, insisting that the law should be formulated as an empirical relation between observed data.

The Belgian astrophysicist and Catholic priest, Georges Lemaître had nevertheless interpreted the radial velocities of spiral nebulae in terms of Einstein's theory in 1927, just two years *after* Hubble had enlarged the Universe, showing that there are extragalactic spiral nebulae, now known as galaxies, which reside outside our Milky Way, but two years *before* Hubble discovered an observed velocity-distance relation for them.

After being ordained a priest in 1923, Abbé Lemaître spent a year at Cambridge University as a student of Arthur Eddington, reviewing Einstein's *General Theory of Relativity*, and during the next two years, Lemaître studied at the Massachusetts Institute of Technology and worked with Harlow Shapley at the nearby Harvard College Observatory.

The Belgian cleric also toured the country, meeting Vesto Slipher at the Lowell Observatory in Flagstaff, Arizona, and Edwin Hubble at the Mount Wilson Observatory, California. As a result, he learned all about the latest measurements of the redshifts, or radial velocities, of spiral nebulae. And this led him to search for an explanation of the observed motions using the *General Theory of Relativity*.

Lemaître spelled it all out in a seminal paper that he published in 1927, the same year that he was made a Professor of Astrophysics at Louvain and earned a Doctorate of Philosophy degree from the Massachusetts Institute of Technology. In his prophetic article, Lemaître interpreted the receding velocities of the extra-galactic nebulae as a cosmic effect of the expansion of the Universe, all in accordance with the relativity theory, additionally deriving a theoretical expression for the linear increase of their velocities with distance.

But we now see this in hindsight, and in 1927 most scientists were not even aware of the observational support for an expanding Universe. Lemaître also published his interpretation in a fairly obscure journal, the *Annales de la Societé Scientifique de Bruxelles*. So practically no one was aware of his findings, which either went unnoticed or were ignored.

At this time, no other astronomer or physicist had used the *General Theory of Relativity* to explain the observed Universe. Einstein adapted it in 1917 to describe a static Universe without motion, while in the same year the Dutch theorist Willem de Sitter found a moving solution without any matter. Eddington summed up the situation at a meeting of England's Royal Astronomical Society in 1930, noting that Einstein's model contains inert matter without motion, while de Sitter's model contains motion but no matter. So the actual Universe, containing both matter and motion, did not correspond to either of these abstract models.

On learning about Eddington's summary, Lemaître sent him a copy of his 1927 paper, reminding him that he had already solved the problem. Eddington realized the significance of his former student's work, and confessed that he had seen it in 1927 but forgotten all about it. Eddington corrected the oversight with a letter to the journal *Nature* that drew attention to the work, and by sponsoring an English translation of Lemaître's paper in the *Monthly Notices of the Royal Astronomical Society* in 1931. In English, the title read "A Homogeneous Universe of Constant Mass and Increasing Radius Accounting for the Radial Velocity of Extra-Galactic Nebulae." Since he was the first to relate the observed galaxy motions to the theory, Lemaître's solution was widely heralded for its novelty, and it led to greater appreciation of the significance of Hubble's discovery.

And if we adopt his explanation, space is swelling, ballooning outward, and carrying the galaxies with it, like birds in currents of air, and the galaxies do not need to be expanding into or through pre-existing space. The galaxies are embedded in space that is being stretched, with the galaxies going along for the ride. So the galaxies are not expanding *in* space, but we instead observe the expansion *of* space.

In this interpretation, the Big Bang was not an explosion that went off in the center of the Universe and hurled matter outward into a pre-existing void. Rather, it was an explosion of space itself that happened everywhere. The wavelength of light is lengthened as it travels in stretched-out space, and the amount of stretching tells us how much the Universe has expanded while the light has been traveling to us.

3.6 AGE, EXTENT AND SHAPE OF THE OBSERVABLE UNIVERSE

How old is the Universe?

Because light travels at a finite speed, to look into space is to look back into time. So powerful telescopes can be used as time machines to see objects as they were in the past. In effect, we are watching cosmic history race at us at the velocity of light, 229,979 kilometers per second. And the look-back time is just the amount of time it takes for light to travel from the object to us at that speed.

It takes 2.6 million years for light to travel from the closest spiral galaxy, Andromeda, to the Earth, and some galaxies are so remote that they emitted the light we see today about 10 billion years ago. These objects are seen as they were then and not as they might be now.

But even on a clear day, you can't see forever. Wherever you look, there is a horizon, and a corresponding limit to the distance between there and now. The cosmic horizon is the look-back time to the Big Bang origin of the expanding Universe, or the expansion age. And that's roughly the reciprocal of Hubble's constant. A low value of the Hubble constant would imply that the Universe is expanding slowly and that the Universe must be relatively old to have reached its current size. By contrast, a high value of Hubble's constant implies a rapid expansion and a relatively young Universe.

Early estimates for the age of the Universe, obtained from the first measurements of Hubble's constant, created quite a problem. Using the value of that constant determined by Hubble and his colleagues, the expansion age only amounted to about 2 billion years. Something wasn't right, for radioactive elements had been used to clock the age of the oldest rocks on the Earth's surface, indicating that some of them are older than the expanding Universe. In 1929, for example, Ernest Rutherford and his colleagues at the Cavendish Laboratory in England worked out the rates of transformations that radioactive uranium undergoes in decaying into lead, and estimated that it would take about 3.4 billion years to produce the amounts of these elements now present in terrestrial rocks.

By the 1950s, Clair Patterson, at the University of Chicago, had made careful measurements of the amounts of uranium and lead locked up in meteorites, obtaining a definitive age of 4.55 billion years, with an uncertainty of just 70 million years. Rounding the number off, we can say, and still do, that everything in the Solar System, from the Earth to the Moon and Sun, is about 4.6 billion years old.

Arthur Eddington wasn't at all troubled by the fact that the Earth seemed older than the expansion age of the Universe, resolving the difficulty by extending the beginning of the Universe further back in time than the start of its expansion. He suspected that the world once existed in an un-moving state described by Einstein's old 1917 model, where a cosmological constant was introduced to produce a cosmic repulsion that opposed the mutual gravitation of all the objects in a static, non-expanding Universe.

According to Eddington's interpretation, in his 1933 book *The Expanding Universe,* the Universe evolved from a stagnating Einstein world of indeterminate age, but then about 2 billion years ago the situation went unstable. The initial non-moving Universe was disrupted by just the slightest disturbance that upset the balance between gravitational attraction and cosmological repulsion. The disturbance caused a slight expansion that thinned out the Universe, making it less able to resist the cosmic repulsion, and the runaway expansion of the galaxies began.

Three young scientists at Cambridge University – Hermann Bondi, Thomas Gold, and Fred Hoyle – developed an alternative Steady-State cosmology in the late 1940s. As they pointed out, you could adjust the cosmological constant to accommodate almost any related observation, and the theory thereby lost its simplicity and uniqueness. So the trio proposed a Universe that had no beginning. The Cosmos, they proclaimed, might have always existed, presenting an unchanging Steady State on the largest scales of space and time. And you would no longer have to attribute the observable Universe to a past creation that was inaccessible to scientific scrutiny or understanding.

Bondi, Gold and Hoyle acknowledged the inescapable fact that the galaxies are moving apart, but supposed that they have always been doing so. Our own Galaxy will never be left alone, they said, because new matter is being created continuously at a rate of just one hydrogen atom per cubic meter of space per year, on average, which is just sufficient to counteract the dispersal and thinning out of the expanding Universe and keep the overall Universe unchanged with time.

This creation of new matter out of the nothingness of space might seem preposterous, but it may be no harder to accept than the supposed creation of matter all at once at the beginning of time.

The age problem that led to serious consideration of the Steady State theory was partially resolved by Walter Baade during World War II. Since he was a German citizen, and therefore classified as an "enemy alien," Baade was restricted to his home in Pasadena, and allowed to leave only to go to Mount Wilson to observe. Most of the other observatory scientists were involved in wartime activities, so Baade had almost unlimited access to the 100-inch telescope during the war, when the lights of Los Angeles and Hollywood were blacked out as a precaution against possible Japanese air raids.

As luck would have it, Eastman Kodak had just developed a red-sensitive emulsion for wartime reconnaissance. Pushing the telescope to its very limits, Baade used the red-sensitive plates to resolve the nucleus of Andromeda, distinguishing individual red giant stars in the crowded center. His measurements, taken in 1943, indicated that these red stars were similar to those found in the globular clusters of our own Galaxy, but different from the highly luminous blue stars found in the outer arms of both the Milky Way and Andromeda.

When the 200-inch telescope on nearby Palomar Mountain began operation, in 1948, Baade continued his investigations of the red and blue kinds of stars, concluding in 1952 that they obey different period-luminosity relations. This meant that Hubble had been confused when applying the distance calibration of Cepheid variable stars in globular clusters to the other kinds of variables in the arms of nearby spiral nebulae. So Baade made the necessary corrections, obtaining a distance of about 2 million light-years for M31, reducing the Hubble constant by about half and enlarging both the scale and age of the expanding Universe by a factor of about two.

Although a consummate observer, Baade tended to avoid detailed analysis and written accounts of his discoveries. So it was fortunate that Henrietta H. Swope, the daughter of the wealthy president of the General Electric Company, joined Baade to assist with the analysis of his excellent 100-inch photographs. Like Henrietta Leavitt, she began her career at Harvard with meticulous analysis of Cepheid variable stars, and continued this work at the Mount Wilson and Palomar Observatories – where she couldn't have obtained her own astronomical observations, since at the time no women were permitted to use the telescopes.

Baade and Swope began a decade-long study of the Cepheid variables in Andromeda, or M31, already concluding in 1955 that their new appraisal of its distance required a downward revision of Hubble's constant to a value of 100, in the usual units. In the meantime, Allan Sandage, continued to correct for Hubble's mistaken identification of the brightest stars, and in 1958 he announced that the elusive constant had a value of 75. Using Sandage's measurements, the expansion time is about 14 billion years, so the Universe was no longer younger than the oldest terrestrial rocks, and scientists were reassured that the Universe did have a definite beginning.

Sandage spent a lifetime measuring the Hubble constant, usually in collaboration with Gustav Tammann of the Universität Basel, Switzerland. By the 1970s, they had reduced the value to about 50, reaffirming it in ensuing decades. During the same period, Gérard de Vaucouleurs, of the University of Texas at Austin, found a large value of Hubble's constant of about 100, using many different and independent methods of determining the distances to galaxies. This would mean that the Universe was only half the size and age that Sandage believed. But the disagreements have narrowed as the result of subsequent research, leading to a value of about 75, between the two extremes and at Sandage's result obtained nearly a half century ago.

Wendy L. Freedman and her colleagues have, for example, used the *Hubble Space Telescope,* abbreviated *HST,* to refine estimates of Hubble's constant and to infer a more accurate age of the Universe. By locating and monitoring Cepheid variable stars in 23 spiral galaxies, members of this *HST* Key Project have directly measured galaxy distances out to 65 million light-years (Fig. 3.15), and connected them to more distant realms of the Universe. In 2001, they used these combined results to infer a Hubble constant of 72 kilometers per second per Megaparsec, with an uncertainty of 10 percent. So the expanding Universe is about 14 billion years old, with an observable extent of 14 billion light-years. If we extrapolate the volume density of nearby galaxies into a sphere whose size is 14 billion light-years, we find that there are roughly a billion galaxies in the observable Universe.

Estimates of the age and size of the observable Universe therefore haven't changed all that much over the past decades. And perhaps the most interesting consequences are that the Cosmos is older than the Earth, and that the light we see from the most distant galaxies was emitted before our Solar System existed.

Clusters of Galaxies

Astronomers have been mapping the distribution of galaxies for more than three-quarters of a century, determining the framework and structure of the Universe. These

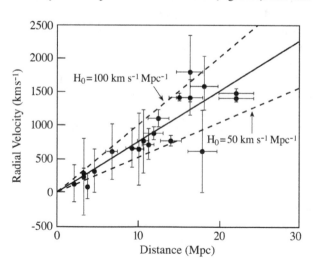

FIG. 3.15   Hubble diagram for galaxies using Cepheid variable stars This plot of galaxy distance versus recession velocity is analogous to that obtained by Edwin Hubble, in his 1929 discovery of the expansion of the Universe (Fig. 3.14). The data shown here summarize eleven years of efforts to measure this constant by using the *Hubble Space Telescope* to measure the distances and velocities of Cepheid variable stars in galaxies. The slope of the linear fit (*solid line*) to the data (*dots*) measures the expansion rate of the Universe, a quantity called the Hubble constant, designated H_0. The distance is in units of a million parsecs, or Mpc, where 1 Mpc is equivalent to 3.26 million light-years, and the radial velocity is given in units of kilometers per second, denoted as km s^{-1}. The fit to these data indicate that $H_0 = 75 \pm 10$ km s^{-1} Mpc^{-1}, and that this constant lies well within the limits of 50 and 100 in the same units (*dashed lines*). [Adapted from Wendy L. Freedman *et al.,* "Final results from the *Hubble Space Telescope* Key Project to Measure the Hubble Constant", *Astrophysical Journal* **553,** 47 – 72 (2001).]

maps determine our place in space; giving us a perspective on the Cosmos and making the far away seem familiar.

Like the initial exploration of the Earth, mapping the Universe has proceeded gradually. First the nearby places were examined, and then the more distant horizons were explored. And eventually the very fabric of the Universe was specified, with maps that reveal its shape, form and texture.

The first cosmic maps were two-dimensional, constructed from meticulous catalogues of the brightest nebulae, as galaxies and some other objects were once called. In 1922, for example, the Swedish astronomer Carl Vilhelm Ludvig "C.V.L." Charlier mapped the distribution of the 12,000 known nebulae that lie outside the plane of the Milky Way (Fig. 3.16), suggesting that cosmic matter is arranged in a hierarchical collection of increasingly larger systems. His maps indicated that the nebulae, now known to be mostly nearby galaxies, are grouped together in hundreds of concentrations, including the one now known as the Virgo cluster of galaxies, named for the constellation that is in the same direction.

During subsequent decades, astronomers compiled larger catalogues that included more numerous fainter galaxies, and an important source for these lists was the photographic plates of the National Geographic Society – Palomar Observatory Sky Survey of the northern sky, completed in 1954 using the wide-angle 1.2-meter (48-inch) Schmidt telescope on Palomar Mountain.

George O. Abell, then a graduate student at the California Institute of Technology, used the newly completed survey to identify clusters of galaxies over a large fraction of the sky, publishing a famous catalogue of 2,712 rich clusters in 1958. Even today, these

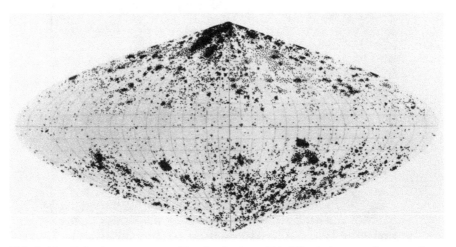

FIG. 3.16 Galaxies congregate The distribution of 11,475 nebulae, now known to be mostly galaxies, plotted in galactic coordinates by the Swedish astronomer Carl Vilhelm Ludvig (C.V.L.) Charlier (1862–1934) in 1922. The bright galaxies group and cluster together in space, forming numerous concentrations. The largest visible galaxy grouping near the north galactic pole (*top*) is known as the Virgo cluster of galaxies. [Reproduced from C. V. L. Charlier, "How an infinite world may be built up", *Arkiv för matematik, astronomi och fysik* **16,** no. 22, 1 (1922).]

dense concentrations of galaxies are referred to as simply "Abell clusters," and designated by the letter "A" followed by the number in his catalogue.

A rich cluster of galaxies typically spans 10 to 20 million light-years, and contains
hundreds and even thousands of individual galaxies (Fig. 3.17). They are held together
by their combined gravity, even though the expansion of the Universe is pulling the
galaxies away from one another.

Thus, instead of being uniformly placed throughout expanding space, like dust
scattered in the wind or a school of frightened fish, the galaxies knot together in great
clusters that are millions of light-years across. In addition, as Abell noticed, the clusters gather and congregate together into larger superclusters, which he called "second-
order clustering." Our Galaxy lies in the outskirts of one of them, known as the Local
Supercluster, which is oriented perpendicular to the Milky Way and extends all the way
to the rich Virgo cluster of galaxies.

FIG. 3.17   Cluster of galaxies The *Hubble Space Telescope* provided
this dramatic view of the center of a massive cluster of galaxies known
as Abell 1689, located 2.2 billion light-years away. The gravity of the
cluster's million million, or trillion, stars, plus any unseen matter, acts
as a gravitational lens in space, bending and magnifying the light of
galaxies located far behind the galaxy cluster, distorting their shapes
and creating multiple images of individual galaxies. (Courtesy of
NASA, the ACS Science team of the HST, STScI, and ESA.)

As time went on and history unfolded, larger telescopes were used to look deeper into space. By counting the number of galaxies in different directions, astronomers created two-dimensional maps similar to Charlier's, but vastly more detailed. They exhibited fascinating, lace-like patterns of connecting galaxies, and curving filaments that enclose dark, seemingly vacant places. This shining web of interconnected galaxies resembled the tangled ribbons of light one sees when looking out from the bottom of a calm ocean bay or a swimming pool.

The massive concentrations of galaxies are using their gravitational pull to locally distort the uniform expansion of the Universe.

Local Eddies, Cosmic Streams

The galaxies are not just flying outward with the expansion of the Universe, in a smooth and regular fashion. Individual galaxies dart here and there, like mosquitoes in a swarm, and entire groups of them are streaming in concert over vast distances, like powerful currents awash in the cosmic sea.

These so-called peculiar motions are caused by the gravitational pull of huge assemblages of galaxies, and they have nothing to do with the uniform expansion of the galaxies, known as the Hubble flow.

For instance, the nearest large galaxy, Andromeda or M31, is moving toward us. Due to the gravitational attraction between the Milky Way and Andromeda, the two galaxies are set on an irrevocable collision course. In 5 or 6 billion years they will meet in a possibly destructive encounter.

And occasionally we catch other galaxies colliding, merging or passing through each other (Fig. 3.18), when they ought to be moving further apart. This mingling is also due to the gravitational attraction of neighboring galaxies, drawing them together and producing local eddies within an outward Hubble flow.

In addition, an entire swarm of galaxies can set off on a trajectory that is independent of the expansion. Because these galaxies are moving together over vast distances of hundreds of millions of light-years, their collective behavior is known as a large-scale streaming motion. It is large in space, but not so big in velocity, generally amounting to no more than 1,000 kilometers per second.

In fact, all the galaxies in our region of space, within a volume of 100 million light-years across, are not just going along for the ride in the pervasive Hubble flow. They are rushing *en masse*, like a high-speed celestial convoy, toward the same remote point in space. In 1987, Alan Dressler, of the Carnegie Observatories, and his colleagues showed that all these galaxies are being pulled through space, forced into mass migration by the gravitational pull of "The Great Attractor." They pinpointed its distance at about 150 million light-years away, and estimated that its mass is equivalent to about 50 million billion (5×10^{16}) stars like the Sun and at least 500 thousand galaxies like the Milky Way. The Great Attractor is most likely a rich and massive cluster of galaxies, part of an even larger supercluster.

Vaster rivers of galaxies might be flowing through more distant regions, as awesome concentrations of mass pull the galaxies here and there in a cosmic tug of war. It all makes you wonder just where we are going, like riding in a taxi with the meter running in a city you've never visited before.

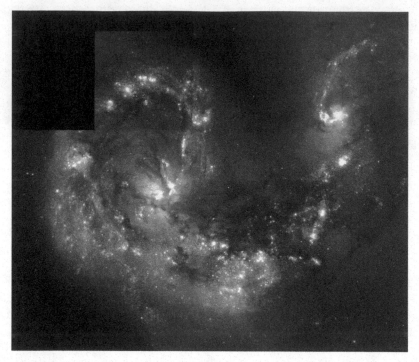

FIG. 3.18   **Colliding galaxies** In this image, taken from the *Hubble Space Tele-scope,* two galaxies, catalogued as NGC 4038 and NGC 4039, are now colliding, spewing trails of newborn stars in their wakes. They are known as the Antennae galaxies, since they look like insect antennae. The galaxies are located 63 million light-years away in the southern constellation Corvus, and have been merging together for about the last 800 million years. As the two galaxies continue to churn together, clouds of gas are shocked and compressed, triggering the birth of new stars that illuminate the long, sweeping tails. The infrared, heat-seeking cameras of the *Spitzer Space Telescope* have additionally revealed star-forming activity in the overlap region where the galaxies meet. (Courtesy of NASA, JPL, Brad Whitmore, STScI.)

So it is the gravitational interaction of galaxies with one another that distorts the smooth cosmic expansion, producing the peculiar motions superposed on the expand-ing Universe. But the uniform Hubble flow gathers speed with distance, like a great river flowing downhill. And since the localized streaming motions are limited in velocity, they are relatively slow when compared with the expansion speed of remote galaxies. The very existence of the large-scale streaming motions nevertheless indicates a decidedly uneven and lumpy distribution of galaxies, which was eventually mapped across billions of light-years.

Galaxies in Three Dimensions

Two-dimensional maps, derived from the catalogued positions of galaxies on the celestial sphere, provide a blurred picture of the galaxy distribution in space. They portray galaxies near and far, piled upon one another at any given location in the sky. Chance superposition of distant and nearby groups of galaxies could therefore be mistaken for real physical structures.

To overcome this problem, astronomers used the radial velocities, or redshifts, of galaxies to gauge their approximate distances, providing the crucial third dimension. At large enough distances, the Universe is ballooning outward with a velocity much larger than the peculiar streaming motions, and the redshifts, measured from the Doppler effect of spectral lines, are assumed to be proportional to distance, all in accordance with Hubble's law. By determining the concentrations of galaxies in different directions and at various redshifts, or depths, the galaxy distribution is brought into sharper focus and greater relief, determining places where the collective force of gravity has pulled galaxies together and locally reversed the uniform expanding motion.

Pioneering redshift measurements in the 1970s and early 1980s indicated that rich galaxy clusters are linked together in elongated superclusters spanning hundreds of millions of light-years. Each supercluster contains about 100 thousand galaxies, with a total mass of about 10 million billion, or 10^{16}, stars like the Sun. In other directions, great voids were found, like the one in the vicinity of the Boötes constellation, discovered in 1981 by Robert Kirshner and his colleagues. These vast, dark empty regions, of perhaps 300 million light-years across, contain very few galaxies and practically nothing else that the telescopes can see.

Moreover, it is sometimes hard to tell where one supercluster ends and another begins, suggesting that they are themselves linked together into enormous filaments that are at least a billion light-years in extent, or about 7 percent the way across the entire observable Universe.

The most startling three-dimensional maps grew out of the decade-long redshift surveys carried out by astronomers at the Harvard-Smithsonian Center for Astrophysics, or CfA for short, located in Cambridge, Massachusetts. Using a modest 1.5-meter (60-inch) mirror mounted in a telescope on Mount Hopkins near Tucson, Arizona, Margaret Geller and John Huchra obtained the first distant, wide-angle, wedge-like map of galaxies, which they called a slice of the Universe.

When these redshift data were displayed in 1986, with the help of University of Paris graduate student Valérie de Lapparent, amazing structures looped into view. The galaxies seemed to be distributed along the peripheries of gigantic hollow bubbles, that nest together in some sort of cosmic foam, like soapsuds in a kitchen sink. Thin elongated strands of galaxies, clusters of galaxies, and the superclusters, swirl across their map in curved structures that resemble the border of wetness left by waves on a beach, or the flowing lines of a Matisse painting. The galaxy clusters appear as bright, concentrated knots and clumps in these strands, whose tangled texture resembles spaghetti alle vongole (spaghetti with small clams), and the long strands delineate the boundaries of huge voids, each seemingly as empty as an immense vacuous bubble.

The CfA redshift survey also delineated an enormous sheet of galaxies, dubbed "The Great Wall", which runs across the middle of the wedge from edge to edge, at distances ranging from 350 to 500 million light-years away (Fig. 3.19). The Great Wall is thin, sharp and well defined, only about 20 million light-years thick despite its great length.

The CfA discoveries motivated three-dimensional mapping projects in the Southern Hemisphere, such as the six-year Las Campanas Redshift Survey from Chile and the four-year Two-degree Field (2dF) Galaxy Redshift Survey from Australia. These cosmic maps also exhibit majestic arcs of shining galaxies that mark the edges of dark voids, hundreds of millions of light-years across.

Modern redshift surveys use electronic technology that permits the simultaneous measurement of hundreds of galaxy redshifts in a single exposure at a large telescope. Optical fibers channel light down to an instrument that records the galaxy spectra using charge-coupled devices to produce a digitized output that is stored within powerful computers. These computers then produce exquisitely accurate and detailed three-dimensional maps of the sky.

The sophisticated new technology has been used to carry out the Sloan Digital Sky Survey, the first comprehensive census of the northern sky since the National Geographic Society – Palomar Observatory photographic survey in the 1950s, but reaching much deeper into the sky with higher precision and more detail. The dedicated telescope, located at Apache Point Observatory in southern New Mexico, has a primary mirror that is 2.5-meters (100-inches) in diameter and an unusually wide field of view, enabling it to map large areas of the sky.

The Sloan Digital Sky Survey is proving once again that the largest features in galaxy maps extend as far as each survey can see. Using the redshifts of just over 11,000 galaxies, the Sloan Survey has reached out into the northern sky to distances of up to 1.5 billion light-years, or roughly a tenth of the radius of the observable Universe. In addition to the expected empty places, or voids, which are present in every survey large enough to contain them, the map reveals the longest sheet of galaxies yet seen. Dubbed the "Sloan Great Wall," it measures 1.37 billion light-years across (also see Fig. 3.19); it is the largest observed structure in the Universe, at least so far.

Thus, everywhere they look, in whatever direction and near or far, modern telescopes are finding a complex and richly textured Universe, filled with luminous concentrations of matter. There is no perceptible end to the lumps and clumps and vacant places, and no one knows where the unevenness will end. Even when looking across 10 percent of the observable Universe, astronomers keep on finding galaxy structures crossing their maps from edge to edge, as well as smaller bubbles, walls and voids that are nestled together in a filigreed pattern.

Terrestrial topographers similarly found new mountains, valleys, continents and seas as they explored distant regions. To those of us who are confined near the surface of the globe, this topography seems rugged, but the tallest mountains reach only one-tenth of one percent, or 0.001, of the Earth's radius. So when you look down on the Earth from a satellite, it is as round and smooth as a basketball, with bumps that are no larger than the dot at the end of this sentence.

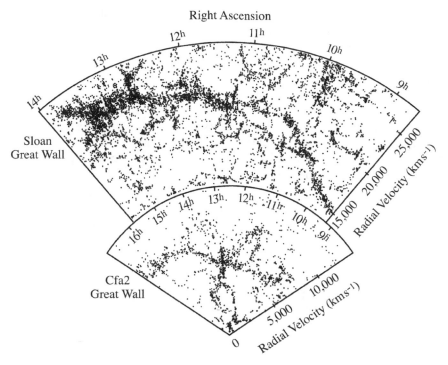

FIG. 3.19 Great walls of galaxies The Center for Astrophysics, abbreviated CfA, redshift survey provided a three-dimensional slice of the distribution of galaxies in the northern sky (*bottom*). The galaxies are concentrated in long, narrow sheets encircling large empty places known as voids, about 100 million light-years in diameter. The dense concentration of dots at the center, forming the body of the "little man", is the Coma cluster of galaxies. The Great Wall of galaxies stretches across the middle of the wedge from edge to edge, across 500 million light-years at radial velocities between 5,000 and 10,000 kilometers per second, abbreviated km s^{-1}. In 2004, the Sloan Digital Sky Survey extended the northern hemisphere survey to radial velocities of up to 28,000 km s^{-1} and an estimated distance of 1.5 billion light-years, discovering a more remote and larger Sloan Great Wall of galaxies that is 1.37 billion light-years long (*top*). It is the largest structure yet observed in the Universe. The Earth is at the apex of both plots, and the radial coordinate, or depth, is the radial velocity, determined from the redshift of spectral lines. Each point on the CfA2 map corresponds to one of 1,732 galaxies, while each dot in the Sloan map corresponds to one of 11,243 galaxies. [Courtesy of Margaret J. Geller, John P. Huchra and the Smithsonian Astrophysical Observatory (*bottom*), and J. Richard Gott and Mario Juric, Princeton University and the Sloan Digital Sky Survey (*top*).]

The lumpiness in the distribution of galaxies, their blemishes and roughness, might similarly blend into apparent uniformity when viewed from a great distance. But it hasn't happened yet. We just can't seem to get far enough away for the unevenness of the Universe to disappear. That remote vantage point is gradually being pushed back to the whole observable Universe, with a radius of about 14 billion light-years.

Yet just about every theoretical speculation about the age, evolution, fate and shape of our Universe assumes that the galaxies have a homogeneous and isotropic distribution, with no preferred place or orientation. Matter, they say, ought to have the same uniform distribution in every direction and at all distances when observed across large enough scales.

Astronomers have indeed found smoothness, when looking back to the beginning of time in the infancy of the Universe, shortly after it began its expansion and before any galaxies existed. They have detected the exceedingly uniform relic radiation of the Big Bang that propelled the expanding Universe into existence. And this brings us to the pervasive, explosive violence of the Universe, ranging from the Earth and Sun to the stars, galaxies and all the way to the Big Bang.

Chapter Four
The Explosive Universe

4.1 PERVASIVE OUTBURSTS

The Spanish artist Pablo Picasso depicted our capacity for destruction in his mural *Guernica* (Fig. 4.1), evoking the brutality, pain and suffering caused by the Nazi bombing of that ancient, defenseless Basque town. The Irish poet William Butler Yeats captured the time, when "things fall apart; the center cannot hold. Mere anarchy is loosed upon the world," and "the ceremony of innocence is drowned; the best lack all conviction, while the worst are full of passionate intensity."[42] And astronomers have recently opened our eyes to previously invisible explosions on a scale hitherto unimagined.

The tranquil abode of steady, unchanging stars and calm, elegant galaxies has been transformed into a realm of cosmic violence. Explosions that are vastly more powerful than any atomic bomb are being detonated on the Sun all the time (Fig. 4.2), hurling lethal particles into interplanetary space and sometimes endangering the Earth. And some stars are blasting themselves apart with unimaginable violence, while gravity compresses their cores into invisibility. Cosmic particles are also being hurled into space from the centers of galaxies and quasars, releasing billions of times as much energy as an exploding star. They are capable of utterly destroying millions of worlds.

In times past, celestial change was viewed as a harbinger of disaster on Earth. Indeed, the word *disaster* once had the meaning of "ill-starred" or "under the influence of an evil star," derived from a combination of the negative *dis* with *astrum*, the Latin word for "star." And the belief that human diseases are influenced by the stars led to the Italian word *influenza*, or the flu for short, which we still use today.

The contemporary lexicon now uses cosmic violence as symbols of power, strength and vitality. There are quasar cars or motorcycles, and pulsar watches. Even the term "nuke them," meaning destruction by a nuclear warhead, is adopted in cavalier expressions of revenge. And then there are the black holes, symbols of the great fear of loss – wealth, health, mind or even life can disappear. Or the most powerful explosion of them all, the Big Bang, which sent the Universe rushing away from itself.

Robinson Jeffers compared this cosmic blast, the Big Bang, to the modern human condition in his poem *The Great Explosion*, which reads in part:

No wonder we are so fascinated with fireworks
And our huge bombs: It is a kind of homesickness perhaps for

FIG. 4.1 **Guernica** This mural by Pablo Picasso (1881–1973) captures the horror of the Nazi aerial bombardment that completely obliterated the defenseless Spanish town of Guernica on 26 April 1927, killing or wounding sixteen hundred civilians. The tortured figures include a menacing, human-faced bull (the fascist General Franco), a gored and speared horse (the Spanish republic), a corpse with wide-open eyes, and a screaming mother cradling a dead child. Arched necks, splayed toes and fingers, and spiked tongues also portray the human agony and suffering. (Courtesy of Museo Nacional Centro de Arte Reina Sofía, Madrid.)

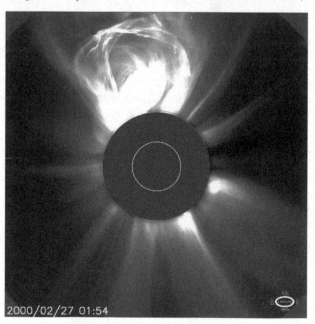

FIG. 4.2 **Coronal mass ejection** A huge coronal mass ejection is seen in this image, taken on 27 February 2000 with a coronagraph on the *SOlar and Heliospheric Observatory,* abbreviated *SOHO.* The white circle denotes the edge of the Sun's visible disk, so this mass ejection is about twice as large as the Sun. The dark area corresponds to the occulting disk of the coronagraph that blocks intense sunlight and permits the overlying solar atmosphere, or corona, to be seen. (Courtesy of the *SOHO* LASCO consortium, *SOHO* is a project of international cooperation between ESA and NASA.)

The howling fireblast that we were born from …
The great explosion is probably only a metaphor – I know not
  – of faceless violence, the root of all things.[43]

Pent-up energy is being released everywhere in the Universe in a kind of cosmic rage, producing endless change. These astronomical discoveries are relatively recent in the historical scheme of things. They could not be fully understood until new technologies widened our range of perception, letting us "see" the unseen by capturing invisible forms of radiation, such as radio waves and X-rays. Telescopes have tuned into these signals and opened new windows onto the Universe, bringing its violent aspects into focus.

4.2 INVISIBLE SKIES

The Electromagnetic Spectrum

Visible light and invisible rays are both known as "electromagnetic" radiation because they propagate by the interplay of electrical and magnetic fields. Electromagnetic waves all travel through empty space at the same constant speed, the velocity of light, never slowing along their long journey and never straying from a straight path. This velocity is roughly 300 thousand kilometers per second, which is a speed that covers a billion kilometers in one hour. And radiation is the fastest thing around. No form of energy or information can be transported more swiftly than the speed of light, a cosmic speed limit enforced by nature's police.

The different forms of electromagnetic radiation are distinguished by their wavelength, the distance between successive crests or successive troughs in the waves. The spectrum of wavelengths ranges from short X-rays on the one end to long radio waves on the other (Fig. 4.3). In the middle of this continuum is the narrow visible slice, covering only about 2 percent of the complete spectrum of all waves emitted by cosmic objects. Radiation with these visible wavelengths, which are about a half-millionth of a meter long, interacts with chemicals in the retina, giving rise to the array of colors that the human eye can see.

The lenses and mirrors that are used to focus and collect visible radiation are described by the science of optics, so the study of visible light from cosmic objects is called optical astronomy. A photograph or electronic picture taken with an optical telescope is called an optical image, and the luminosity of a cosmic object at visible wavelengths is known as its optical luminosity. And an optical identification establishes the visible-light counterpart of an object detected in some other part of the electromagnetic spectrum.

Astronomers now tune into the invisible Universe, which lies well beyond the range of visual perception, by observing cosmic radio waves, with wavelengths as big as boulders, houses or even mountains, and cosmic X-rays that are as small as an atom. Radio telescopes collect radio waves and provide one perspective; they are the tools of radio astronomy. And X-ray telescopes gather in cosmic X-rays, providing a different view within the domain of X-ray astronomy.

Radio waves cover an enormous range of wavelengths between 0.001 and 100 meters, and all except the longest ones pass easily through our atmosphere. They are even unperturbed when passing through stormy weather, and that is the reason your car radio works when it rains or snows outside.

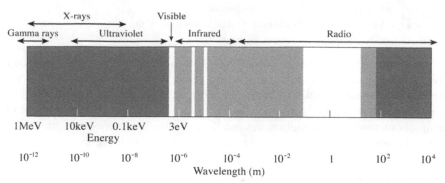

FIG. 4.3 Electromagnetic spectrum Radiation from cosmic objects can be emitted at wavelengths from less than 10^{-12} meters to greater than 10^4 meters. The visible spectrum that we see with our eyes is a very small portion of the entire range of wavelengths. Lighter shading indicates a greater transparency of the Earth's atmosphere. Cosmic radiation only penetrates to the Earth's surface at visible and radio wavelengths, respectively represented by the narrow and broad white areas. Electromagnetic radiation at short gamma-ray, X-ray and ultraviolet wavelengths, represented by the dark areas, is absorbed in our atmosphere, so the Cosmos is now observed in these spectral regions from above the atmosphere in Earth-orbiting satellites.

The wavelengths of the longer radio waves are difficult to measure, while the frequency is more easily determined. The frequency of a wave is the number of crests passing a stationary observer each second; the frequency therefore tells us how fast the radiation oscillates, or moves up and down. The product of wavelength and frequency equals the velocity of light, so when the wavelength increases the frequency decreases and *vice versa*.

Radio stations are denoted by their call letters and the frequency of their broadcasts, usually in units of million of cycles per second, or MegaHertz abbreviated MHz, for frequency modulated, or FM for short, broadcasts. One complete vibration, or cycle, per second is called a Hertz, named after the German physicist Heinrich Hertz who discovered radio waves in 1887.

The main reflector of a radio telescope is known as a dish because of its parabolic dish-like shape. The curved metal surface gathers the incoming radio waves, reflecting and focusing them to an electronic receiver. This receiving system converts the intensity of the incoming radio signal to numbers that are manipulated in a computer to form images.

But since radio waves are millions of times longer than those of light, a radio telescope needs to be at least a million times bigger than an optical telescope to obtain the same resolving power. At a wavelength of 0.5 meters, for example, the reflector of a radio telescope needs to be almost 2 kilometers across to achieve the same resolution as that of the unaided human eye, about a minute of arc at visible wavelengths. All single radio telescopes are much smaller than this, so by themselves they provide a very myopic, out-of-focus view, comparable to looking at the radio Universe through the bottom of a glass bottle. Radio astronomers get around this difficulty by connecting smaller radio telescopes electronically, simulating a larger one through the technique of inter-ferometry, which analyses how waves detected at the smaller telescopes interfere once they are added together – hence the term *interferometer* for "interference meter."

X-rays lie at the short-wavelength side of the electromagnetic spectrum, extending from a wavelength of one-hundred-billionth (10^{-11}) of a meter, which is about the size of an atom, to the ultraviolet edge of the visible. The X-ray wavelengths are so short that it is impractical to measure them directly, and it is more convenient to describe this form of radiation by its energy.

Electromagnetic radiation is a wave when it travels through empty space, but it becomes a particle when it encounters the material world. That is, whenever radiation is absorbed or emitted by atoms, it behaves not as a wave but as a package of energy, or a particle. These packages are given the name photons. They are created whenever a material object emits electromagnetic radiation, and they are consumed when matter absorbs radiation.

The ability of radiation to interact with matter is determined by the energy of its photons. Radiation with shorter wavelengths has greater energy. This is the reason that short-wavelength, energetic X-rays pass through your skin and muscles with very little absorption, until they encounter the dense bones beneath. In contrast, longer-wavelength, less-energetic visible light just warms your face and your skin is all we see.

The unit of energy that is used in X-ray astronomy, the electron volt, is related to the way X-rays have been generated since 1895, when the German physicist Wilhelm Röntgen discovered them. To produce X-rays you just connect a high-voltage power supply to two metal electrodes placed inside an evacuated glass tube, generating an electrical current that passes between them. The high-voltage electrons emit X-rays when they strike the target metal electrode, and are deflected by atomic nuclei there. One electron volt, abbreviated eV, is the energy an electron gains when it passes across the terminals of a 1-volt battery, and one thousand electron volts, or 1 keV with "k" for kilo or one thousand, is the kinetic energy of an electron that crosses a potential difference of 1,000 volts.

Cosmic X-rays are totally absorbed in the Earth's atmosphere, so they must be observed using telescopes lofted above the air. And although visible light is reflected by mirrors, incident X-rays are absorbed in them. The X-rays can nevertheless glance off incredibly smooth, metal surfaces at low angles, like a flat pebble skimmed over a pond. Many X-ray telescopes launched into space therefore consist of highly polished, slightly tapered, rounded and nested sheets of metal.

X-ray astronomers like to distinguish between soft and hard X-rays. The soft X-rays have modest penetrating power with relatively low energies between 1 and 10 keV. Hard X-rays have greater penetrating power and higher energies, lying between 20 and 100 keV. As a metaphor, one thinks of the large, pliant softballs and the compact, firm hardballs, used in the two kinds of American baseball games.

Gamma rays are the most energetic form of electromagnetic radiation, similar to X-rays but with even greater photon energy. As you might expect, gamma rays have ener-gies associated with nuclear processes, which occur within the nuclei of atoms or during nuclear interaction. Their energy lies between 100 thousand and one million electron volts, respectively designated 100 keV and 1 MeV, with "M" for mega or a million.

Discovery of the Invisible Radio Universe

The American radio engineer Karl Jansky accidentally discovered cosmic radio signals in the early 1930s when radio waves were being used extensively for global communications, and nobody was listening to anything outside the Earth. At that time the Bell Telephone Laboratories, or Bell Labs for short, assigned Jansky the task of tracking down and identifying natural sources of radio noise that were interfering with shortwave radio communications. Shortwave meant, in this case, wavelengths near 14.6 meters (20.5 MHz) that were used for ship-to-shore radio communications.

In addition to the noisy static produced by terrestrial sources, such as lightning discharges from both local and distant thunderstorms, and the inner workings of the radio receivers themselves, Jansky detected a persistent and faint hissing noise, somewhat like an angry cat. The maximum intensity occurred four minutes earlier each day, and a friend told Jansky that this would be expected if the source was celestial, rising and setting from a fixed point in the sky like a star observed from the rotating Earth.

Using a novel rotating, directional antenna system for more than a year, at the Bell Labs field station at Holmdel, New Jersey, Jansky established the general direction and arrival time of the most intense radio waves, eliminating the Sun as a source and showing that the radiation came from outside the Solar System.

The serendipitous discovery was reported by newspapers throughout the world, including the *New York Times* whose front-page headline for 5 May 1933 read "New Radio Waves Traced to the Center of the Milky Way". Jansky even appeared on a radio program, which rebroadcast his "star noise" so listeners could hear "the hiss of the Universe" by a direct long-line connection from Holmdel to the New York broadcasting studio.

And the popular press was right; this was a very important discovery. Cosmic radio waves tune into worlds that cannot be seen in visible light, enabling us to view unique aspects of the Universe during night or day and rain or shine. It is like standing by the shore of a lake on a moonless night. Only the sound of moving waves tells you that something is out there.

In the 1930s, the country was in the throes of the Great Depression, when jobs were scarce and the specific communications interests of the Bell Labs were of more importance to Jansky's employment. So he didn't return to astronomical research after 1935, and he never received any scientific award for his profound discovery of extraterrestrial radio waves.

Moreover, the astronomical community almost completely ignored the result. Jansky did not publish in an astronomical journal, and his radio techniques were so much outside the conventional methods of astronomy that no traditional observatory contributed to new knowledge about it. So astronomers did not become aware of the different way of looking at the Cosmos until the 1940s, when a radio engineer and amateur astronomer, Grote Reber, confirmed and extended Jansky's findings.

With the occasional help of a local blacksmith, Reber built with his own hands a 7.6-meter (30-foot) parabolic radio telescope in the backyard of his home in Wheaton, Illinois. Reasoning that the cosmic radio sources would emit most of their energy at shorter wavelengths, as the thermal radiation of any hot gas will, he started his observations at relatively short radio wavelengths, but when nothing was detected he tuned his receiver to longer waves. And after a few years of trying, Reber detected radio noise coming from the Milky Way at a wavelength of 1.87 meters. So he concluded that the radiation must increase in intensity at longer wavelengths rather than shorter ones.

Reber discussed these results with the astronomers at the Yerkes Observatory, and convinced Otto Struve, the editor of the *Astrophysical Journal,* to publish his article on "Cosmic Static" in 1940, thereby exposing the results to the astronomical community. A second paper, presenting complete maps, was also published in the *Astrophysical Journal,* in 1944. The contour maps showed the general, diffuse radio emission from the Milky Way, and the intense source at the galactic center in the direction of the constellation Sagittarius. They also revealed prominent, localized patches of intense radio emission at a few other places on the sky, including ones located in the directions of the Cassiopeia and Cygnus constellations.

When first discovered, the most intense radio sources were given the name of the constellations they happened to be in the direction of, followed by the letter A for the first and strongest source found in that direction. So the discrete sources detected by Reber were designated Sagittarius A, Cassiopeia A, and Cygnus A.

World War II (1939 – 1945) spurred the development of radio detection and ranging, abbreviated radar. It had the magical ability to "see" things at a great distance, through clouds or darkness and during day or night, by bouncing powerful radio pulses off them and detecting the "echoes." Allied fighter airplanes carried radar transmitters and detectors aloft, enabling them to locate enemy aircraft in the dark and to detect enemy ships or submarines that had risen to the surface of the sea at night.

Finding themselves with laboratories full of equipment, and no enemy in site at the end of the war, radio physicists and engineers applied their wartime experience with radio technology and radar development to pioneer radio astronomy. This occurred primarily in two countries, Australia and England.

The Australians set up a defense radar antenna on a cliff overlooking the sea, and combined the rays coming directly from an object with those reflected by the mirror-like water. The technique was an early example of an interferometer, since it analyzed how the direct and reflected waves interfered once they were added together.

The cliff, or sea, interferometer provided sufficient angular resolution to identify three discrete radio sources with known astronomical objects in 1949. They were: Taurus A – the Crab Nebula supernova remnant, Virgo A – a giant elliptical galaxy designated as M87, with a jet of light extending from it, and Centaurus A – an unusual galaxy, denoted NGC 5128, with a dark lane of dust across its center (Fig. 4.4).

In England, at the Cavendish Laboratory of Cambridge University, Martin Ryle constructed a spread out "array" of modest-sized radio telescopes, connecting them electronically to make several interferometer pairs and simulate a single larger radio telescope. This improved both the angular resolution of intense radio sources

FIG. 4.4   Centaurus A One of the first extragalactic radio sources to be discovered, Centaurus A, is associated with the elliptical galaxy NGC 5128. It lies at a distance of about 15 million

light-years, and is the nearest radio galaxy. This composite image shows the optically visible component (*orange and yellow*), radio (*pink and green*), and X-ray (*blue*) emission. A broad band of dust and cold gas crosses the visible galaxy. The radio emission of Centaurus A is concentrated within two lobes, which is characteristic of most radio galaxies. In this case, the radio-emitting lobes jut out at right angles to the dust lane, extending across seven degrees of sky, equivalent to about one million light-years. The radio lobes are fed by jets of high-energy particles that were probably ejected from a super-massive black hole in the nucleus of the optically visible galaxy. Two large arcs of X-ray emitting gas, heated to millions of kelvin, may have been produced in a titanic explosion about ten million years ago. [Courtesy of Digitized Sky Survey U.K. Schmidt Image/StScI (*optical*), NRAO/AUI/NSF/VLA (*radio*), and NASA CXC (*X-ray*).]

and the sensitivity needed to detect weaker ones. As a cosmic radio source moves across the sky from horizon to horizon, the effective antenna separations change, and the results at varying resolution can be combined. The Nobel Prize in Physics was awarded to Ryle in 1974, primarily for his development of this technique of aperture synthesis.

The true nature of discrete radio sources was not immediately apparent. In 1951, for example, Ryle and his colleagues had found that the fifty most intense sources have a uniform distribution across the sky, with no preferred directions and no concentration in the Milky Way where most stars are found. At a debate over the origin of cosmic radio noise, held that year at the University College, London, Ryle emphatically insisted that a new kind of radio star had been discovered. The imaginative Thomas Gold, who was then lecturing in physics at Cambridge University, disagreed, reasoning that the discrete radio sources are not concentrated in the Milky Way because they have nothing to do with our Galaxy, and are located much further away.

The controversy was settled when English radio astronomers, both at Cambridge and at the University of Manchester, used radio interferometer techniques to refine the position of the intense radio source Cygnus A, setting the stage for the crucial optical identification that would transform our knowledge of the radio Universe. Armed with the accurate radio position, two German-born American astronomers, Walter Baade and Rudolph Minkowski, used the giant 5.0-meter (200-inch) optical telescope on Mount Palomar, California to find the visible-light counterpart of Cygnus A in 1954. It was an elliptical galaxy with a redshift of 0.057, which meant that it is receding from us at 5.7 percent of the velocity of light. And its distance, of about 750 million light-years, was inferred from the linear relationship between redshift and distance, known as Hubble's law.

Once this distance had been established, the enormous absolute radio luminosity of Cygnus A was realized. It wasn't just emitting a faint crackle and hiss, but instead a colossal, shattering roar, like a lion instead of a household cat. Every second, Cygnus A emits as much energy in radio waves as a million million, or 10^{12}, Sun-like stars radiate in visible starlight. It turned out to be a new type of radio galaxy, whose radio luminosity is comparable to its optical one.

The news that Cygnus A was a galaxy, and not a star, embarrassed and humiliated Ryle. When he saw the identifying photographs, at a scientific conference, he threw himself on a nearby couch – face down, buried in his hands – and wept.[44] But he soon recovered his composure, realizing that the radio galaxies could be used to probe the distant Universe and provide tests of different cosmological models.

And Ryle's group did just that, abandoning the radio star hypothesis and using their comprehensive and definitive Cambridge radio surveys to demonstrate strong evolutionary effects over cosmic times. The number counts of the fainter, presumably more distant, radio sources differed from those of the strong nearby ones, which was interpreted as an evolutionary effect in which the more distant, younger radio galaxies are either more luminous or more concentrated than the nearby older ones.

As Ryle noticed, evolving radio sources contradicted the Steady State cosmology advanced by Fred Hoyle, Hermann Bondi, and Thomas Gold, which assumes that the Universe is the same at all times and places. As you look out into space, and thus back in time, the radio galaxies should resemble each other, but they do not. This discrepancy helped overthrow the Steady State cosmology.

Discovery of the Invisible X-ray Universe

The other new window on the explosive Universe was opened at the opposite end of the electromagnetic spectrum, in short-wavelength X-rays. They illuminate the most energetic features of the Universe – the hottest gases, the strongest gravity, the most intense magnetic fields, and the most energetic explosions. And like radio astronomy, X-ray astronomy received its impetus from wartime technologies and expertise.

The main motivation for the very first cosmic X-ray observations was to understand the Sun's control of the Earth's ionosphere. Global radio communications, made possible by reflection off this ionized layer, were sometimes disrupted by solar activity, and the military wanted to understand why. So in the 1950s, Herbert Friedman and his colleagues at the United States Naval Research Laboratory used instruments aboard captured German V-2 rockets, and subsequently their own sounding rockets, to show that X-rays from the Sun ionize most of the upper atmosphere.

They also showed that the solar X-rays rise and fall with the number of sunspots, and that the X-ray radiation emitted during a solar explosion can outshine the entire Sun at these wavelengths. The apparently serene Sun, an unchanging disk of brilliant light to the naked eye, has no permanent features in X-rays, which describe a volatile, unseen world of perpetual change.

If the Sun was any guide, then X-rays could not be observed from any other star. Even the nearest stars would be too far away, and their X-ray emission too faint to be detected with existing instruments. The fortuitous discovery of X-rays from cosmic sources other than the Sun therefore came about unexpectedly, when Riccardo Giacconi and his colleagues at the American Science and Engineering Company, abbreviated AS&E, placed sensitive Geiger counters on a rocket to search for possible X-rays emitted when energetic solar particles strike the Moon, as well as for other unknown cosmic sources that might have unforeseen methods of producing X-rays.

The Geiger counters aboard a five-minute AS&E rocket flight, launched from White Sands, New Mexico in 1962, detected discrete sources located far from the Moon or Sun. The strongest signal was measured when the detectors were pointed toward the constellation Scorpius, and a second, significantly weaker source appeared in the direction of Cygnus. Such intense X-ray radiation was totally unanticipated, for it was up to a hundred million times brighter than that expected from even the nearest star other than the Sun.

The most intense X-ray sources have been named after the constellation they appear in, followed by an X for X-ray and a number ordered by decreasing brightness. Famous examples include Cygnus X-1, Cygnus X-3, Centaurus X-3, Hercules X-1 and Scorpius X-1. The numerous fainter sources are usually named by a catalogue designation followed by the source's position in the sky – its right ascension and declination.

After Giacconi's pioneering rocket flight, NASA took over as the major sponsor of his group, including the development of the first dedicated X-ray satellite observatory to be flown as part of NASA's *Explorer* program. That satellite was launched on 12 December 1970 from the Italian site of San Marco, off the Coast of Kenya. And since this date coincided with the seventh anniversary of the independence of Kenya, the satellite was named *Uhuru*, the Swahili word for freedom.

Due to the great sensitivity of its X-ray detectors, *Uhuru* found many sources that are less intense than the first ones to be discovered, while also observing the strongest sources with high time resolution. It raised the count of known X-ray sources to 400, and discovered rapid, irregular variations in the X-ray intensity of Cygnus X-1 on times scales as short as 0.1 seconds. This meant that Cygnus X-1 had to be an extremely compact object. Another newly discovered source, Hercules X-1, emitted strong, periodic X-ray bursts every 1.2378 seconds. The Nobel Prize in Physics was incidentally awarded to Giacconi in 2002 for these pioneering contributions to astrophysics, which led to the discovery of cosmic X-ray sources.

The X-ray emitting particles in these objects have been heated to very high, million-degree temperatures as the result of intense gravity. Such conditions can exist near collapsed, dead stars, including black holes and neutron stars, which Cygnus X-1 and Hercules X-1 respectively turned out to be. So X-ray telescopes can reveal the skeletons of the Cosmos, much as X-rays detect the bones of humans. X-ray astronomy has therefore brought dead stars back to life. They are lit up by gas supplied by living visible companions.

An understanding of their unseen partners, the black holes and neutron stars, includes a realization that the cores of dead stars, which have used up their thermonuclear fuel, collapse while their outer stellar layers are hurled outward in powerful explosions.

4.3 STARS THAT BLOW THEMSELVES APART

Guest Stars

For at least two thousand years, astronomers, hunters, mariners and others familiar with the brightest stars must have been amazed by a *nova*, or "new star," that would appear suddenly at a point in the sky where no star had been previously seen. For a few days, the nova might be among the brightest stars in the dark night sky. But then the star would begin to fade away, and in about a month it would disappear back into invisibility, without a trace.

The nova might be amongst the brightest stars for just a few days, or shine as brightly as a planet for a year or two and then disappear from view. The Chinese called them "guest stars" or "visiting stars" since they were not permanent members of the celestial sphere, instead appearing suddenly and then departing abruptly like uninvited guests.

Each Chinese ruler was believed to be the Son of Heaven, able to rule only so long as he fulfilled Heaven's wishes. The Chinese astronomers were therefore employed by their emperor to routinely watch the heavens for signs of their influence on the destinies of men. Conspicuous changes in the sky were taken very seriously – according to legend two Chinese astronomers were beheaded for failing to predict a solar eclipse about four thousand years ago.

Sometimes the new star would become so bright that it was easily visible even in full daylight. The emperor's astronomers in the Sung dynasty of China recorded one of them on 4 July 1054, near the constellation now known as Taurus, the Bull. The Chinese chronicles indicate that the new star became as bright as Venus, could be seen during the daytime for three weeks, and remained visible in the night sky for 22 months.

More than four centuries passed before the Oriental records again noted the unheralded appearance of brilliant guest stars, and this time they shook the very foundations of European thought. As Aristotle taught, heavenly bodies were supposed to be eternal, pure, changeless, incorruptible and perfect, quite unlike anything on Earth. Yet, in a span of just 32 years two new stars appeared, remained fixed in the heavens for about a year each, and then disappeared from view. Tycho Brahe witnessed the first visitor in 1572; it appeared in the constellation Cassiopeia with a light more luminous than the planet Venus. Then Johannes Kepler spied another one as it lit up the heavens in 1604.

Light of a Billion Suns

In the early 20th century, astronomers discovered that there are two classes of new stars that can be seen using large telescopes, the common faint novae that fade rapidly, and the more luminous, long-lasting type that are a million times more energetic. The Swiss astronomer Fritz Zwicky gave the brightest kind the name super-novae – or supernovae in the modern de-hyphenated form. The guest stars of 1054, 1572, and 1604 are examples of supernovae.

Eccentric, aggressive and independent Professor Zwicky had the dangerous habit of telling everyone just what he thought of him or her. He was always out "to show those bastards", often meaning his colleagues at the California Institute of Technology, or Caltech for short, and once called them "spherical bastards," because, he said, "they were bastards any way you looked at them."

Such abrasive and outspoken behavior didn't endear him to his fellow humans, but Zwicky didn't care – he knew he was smarter than most of the Philistines he had to work with. And his quick intelligence and irascible personality made him a formidable and indefatigable interrogator of German scientists after the war, including Wernher von Braun and others associated with the German V-2 rockets. To Zwicky, their V-2 achievement was due less to the accomplishment of individual genius, then to the work of a large number of "moderately technical individuals", and he might have added slave labor to the description.

In 1934 Walter Baade and Zwicky communicated to the United States National Academy of Sciences a remarkable pair of papers on supernovae. In one paper, they demonstrated that the enormous total energy emitted in the supernova process corresponds to the complete annihilation of an appreciable fraction of the star's mass. As we now know, supernovae are stars that end their life with a bang, collapsing inside and blowing themselves apart on the outside, briefly shining with the light of a billion stars.

The other, more speculative paper by Baade and Zwicky discussed their prediction that a supernova explosion accelerates charged particles to high energies, most likely accounting for the energetic, cosmic-ray particles that rain down on the Earth from all directions. If supernovae occur once in a millennium in our Galaxy, they can account for the observed flux of cosmic rays. Almost as an afterthought, included under "Additional Remarks," they added that the imploding core of a supernova explosion represents the transition of an ordinary star into a neutron star, consisting only of neutrons.

It took half a century for astronomers to realize that Baade and Zwicky were right on all counts. Then, with his characteristic modesty, Fritz claimed to have successfully made the most concise triple prediction ever made in science.

In the meantime, World War II (1939 – 1945) came along. Zwicky incorrectly accused Baade, who was a German "alien", of being a Nazi, and threatened to kill him if he showed up on the Caltech campus. From then on, Baade refused to be left alone in a room with Zwicky, who used to entertain visitors by doing one-armed pushups.

What makes an isolated star blow itself apart, self-destructing in a cataclysmic, one-time act of stellar suicide? The bright explosion is a symptom of old age, the last convulsions of a massive star that has used up all its internal nuclear energy. Deprived of these resources, the star collapses under its own weight, and in the aftermath an explosion occurs. Thus, what were once thought to be exceptionally bright new stars are really catastrophic outbursts from very old, massive stars.

At the end of its bright life, with its nuclear fires spent and the heavy hand of gravity bearing down, a dying star emits one last spurt of energy. It reminds one of the Welsh poet Dylan Thomas, who wrote, "do not go gentle into that good night, old age should burn and rave at close of day; rage, rage against the dying of the light."

All stars begin their lives by burning hydrogen into helium, but the more massive stars have a shorter life expectancy, like some overweight people. And when the core hydrogen is expended in such a massive star, the star's core will contract until it becomes hot enough to burn helium, and the stellar envelope will expand into a supergiant star. Aging then accelerates and progresses rapidly, as it can for humans. The bloated giant quickly consumes its internal fuel sources at ever-increasing central temperatures and rates, until there is nothing left in the core but pure iron. All the available nuclear fuel has been exhausted, and the star has reached the end of the line.

When the iron nuclei in the core of such a star are pushed together, no energy is released. The iron does not burn, no matter how hot the star's core becomes, and the nuclear fires are extinguished. The star has become bankrupt, completely spending all its internal resources, and there is nothing left to do but collapse. That is, there is no longer any energy being generated to sustain the star's structure and keep it pumped up against the force of gravity.

Then a massive star, having burned brightly for perhaps ten million years, can no longer support its own crushing weight. The center collapses in less than a second, accruing energy from its in-fall, much the way a hammer or waterfall gains energy when moving toward the ground.

The central iron core is crushed into a ball no bigger than New York City. Electrons are squeezed inside the iron nuclei, combining with their protons to make neutrons. The material is compacted to nuclear density, and the center collapses to form a neutron star.

Having lost the supporting core, the surrounding material at first plunges in toward the center, like a building with the foundation removed. But the core, rushing inward faster than the envelope, bounces back, and the outer collapsing layers encounter the blast of the rebounding core, like a bullet hitting a brick wall. The pent-up energy of the central core tears the outer layers of the star apart, expelling them into deep space at supersonic speeds. The doomed star suddenly increases in brightness a hundred million fold, becoming a supernova that briefly outshines up to a billion of its neighboring stars put together. Without this supernova explosion, many of the heavy elements now found in the Cosmos would remain forever trapped within dead stars.

But there is more than one way to explode a star. The scenario we have just described applies to an isolated star with the right mass, between about 8 and 20 times the Sun's mass. When that sort of massive, dying star exhausts the nuclear power source at its center, the core collapses into a neutron star in less than a second, and an explosion blows away all the rest of the in-falling matter. This kind of gravity-powered supernova explosion of isolated, massive stars is known as type II, to distinguish it from a different sort of thermonuclear explosions, labeled type Ia (Focus 4.1).

Thermonuclear supernovae of type Ia and gravity-powered ones of type II are believed to occur with about equal likelihood in a galaxy, at the rate of one every 50 to 100 years. Light from hundreds of them might be on the way to us now within our own Galaxy, the Milky Way. Their light could take many thousands of years to travel to us. We can now also use telescopes to see supernovae that occurred billions of year ago in far away galaxies. And not very long ago, one of them blew up in a nearby, irregular galaxy with enough brilliance for you to see it, if you lived south of the Equator and happened to look up at night.

FOCUS 4.1
Supernova Explosions of Another Type

In 1960, the English astrophysicist Fred Hoyle and his American colleague William A. "Willy" Fowler proposed a new kind of supernova, now known as type Ia and characterized by the absence of emission from hydrogen. It occurs in a close binary system, with a white dwarf, the shrunken dense remnant of an ordinary low-mass star, circling another nearby ordinary star. The white dwarf is an Earth-sized star that is not massive enough to ignite its unburned thermonuclear fuel. So it's a potential bomb that remains harmless unless detonated, like a stick of dynamite.

When a nearby companion stars expands, as the result of its normal evolution, hydrogen from its outer atmosphere spills onto the white dwarf. The overflow might, for example, happen when the ordinary visible companion runs out of hydrogen fuel in its core,

and swells into a red giant star. An explosive situation has developed, and the mass transfer can push the white dwarf over the edge of stability by increasing its mass, triggering runaway, explosive nuclear reactions that quickly spread throughout the star and completely shatter it.

Since every one of these explosions is triggered at the same mass limit, under very similar conditions, and also since the star is completely destroyed, type Ia supernovae are expected to produce about the same maximum light output every time they occur. Astronomers find this bright, uniform luminosity very useful as a "standard candle" used to measure the distances of very remote galaxies, and to thereby determine the pace of cosmic expansion. In contrast, type II supernovae, of single, massive stars without a close companion, vary quite a bit in their intrinsic brightness, and this makes them less useful for gauging distances.

SN 1987A

Late in the evening of 24 February 1987, astronomers discovered the first supernova seen with the naked, or unaided, eye since 1604. The type II supernova, dubbed SN 1987A, exploded 166 thousand years before that night, generating an intense burst of light in the Large Magellanic Cloud, one of the two satellite galaxies of the Milky Way visible from the Earth's southern hemisphere.

Astronomers at the Las Campanas Observatory, on a barren mountaintop near La Serena Chile, were the first to notice the new star. Ian Shelton photographed it using a small 0.25-meter (10-inch) telescope during a routine survey of the Large Magellanic Cloud, and at roughly the same time Oscar Duhalde, an operator of a nearby 1-meter (40-inch) telescope, independently saw the supernova with his own eyes, when stepping outside to check the clarity of the skies. Halfway around the globe in New Zealand, an amateur astronomer, Albert Jones, noticed the supernova just four hours later. It was by this margin that Shelten became the discoverer of SN 1987A, the first supernova to be detected with or without a telescope that year.

The discovery was relayed to the International Astronomical Union's clearinghouse for such events, which telegraphed the news to astronomers throughout the world. By the next evening, nearly all major radio and optical telescopes south of the Equator were observing the supernova. One month later, it was featured on the cover of *Time* magazine with just one word: "BANG!" And astronomers were still watching its expanding debris years after the exploding star hurled it into space (Fig. 4.5).

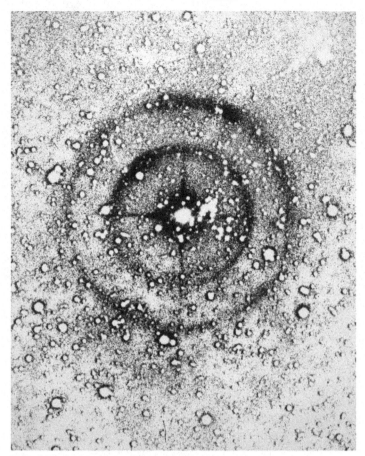

FIG. 4.5   Light echoes from SN 1987A Two complete rings of
light surround the exploded star SN 1987A in this negative image
taken with the 3.9-meter (153.5-inch) Anglo-Australian Telescope on
15 July 1988, a little more than a year after the supernova was first
sighted. These light echoes arise in two thin sheets of material located
about 470 light-years (*inner ring*) and 1,300 light-years (*outer ring*) in
front of the supernova. Microscopic dust grains in these sheets have
been illuminated by light that left the supernova when it was brightest, in mid-May 1987. The rings have been made more prominent by
photographically subtracting an image taken three years before the
supernova exploded, canceling much that existed previously. Stars,
however, are still visible as faint haloes. (Courtesy of David Malin and
the Anglo-Australian Observatory.)

The supernova explosion SN 1987A even emitted neutrinos, which were recorded
hours before the first visible sighting in two neutrino detectors buried deep underground. These neutrinos are close to being nothing at all. They move very near the
velocity of light, have no electrical charge, and so little mass that practically nothing
gets in their way. They are the true ghost riders of the Universe, traveling almost unim-

peded through nearly any amount of matter. So they are very hard to detect. Massive, underground neutrino observatories only manage to snare a few of the trillions of neutrinos passing through them.

In just a few seconds, two subterranean neutrino detectors together recorded just 20 neutrinos flashing through the underground darkness. Yet, they signaled the presence of an awesome power, vastly exceeding the energy contained in the light and expanding debris of SN 1987A. By measuring the amount of energy of the detected neutrinos, scientists were able to show that the supernova generated 10^{58} neutrinos in just ten seconds, emitting a total neutrino energy that briefly exceeded the luminous output of all the stars of every galaxy in the Universe!

The neutrinos detected from SN 1987A solved one of the thornier problems in understanding supernova explosions. It was known that the stellar collapse generates tremendous amounts of energy, but there was trouble explaining how that energy was transferred from the core into the outer layers in sufficient amounts to produce an explosion. The collapsing core might rebound, sending shock waves propagating into the surrounding material, but computer simulations indicated that the shock waves could not blow the rest of the star away. They always got stalled when encountering the in-falling matter from the outer layers, like the mythological Sisyphus who just can't push that rock over the hill or a car trying to get up an icy slope during a winter storm.

Unlike the stalled shock waves, the flood of escaping neutrinos carries tremendous amounts of energy far away from the stellar core, and a very small fraction of them get caught in the in-falling outer layers, heating them up to a temperature of more than 10 billion kelvin. This produces a buoyant, convecting bubble of energy that reverses the in-fall and powers the explosion.

With the help of the neutrinos, the shocks are revived, and they can break through the confining outer parts of the star. As Captain Ahab said in Herman Melville's *Moby Dick:* "How can the prisoner reach outside except by thrusting through the wall." Nowadays, we might also be reminded of the song by the rock group *The Doors* entitled "Break on through to the other side."

Three hours after the initial collapse and generation of neutrinos in SN 1987A, its heated bubble expanded, driving shock waves before it, and burst through the surrounding material, breaking the star apart, hurling its pieces into space and producing the dazzling light that signaled the birth of the supernova. That explains why its neutrinos were detected hours before any light was seen.

Expanding Remnants of Shattered Stars

We can still perceive the debris of cataclysmic stellar explosions that occurred in our Galaxy thousands of years ago, often before recorded history. Radio astronomers have found more than 150 of these expanding supernova remnants, and X-ray astronomers have observed the high temperature gas ejected by many of them.

The most spectacular example of such a supernova remnant is the Crab Nebula (Fig. 4.6). As a New Orleans' drummer might have said of a beautiful, spirited jazz singer, "she has everything." The Crab is the first nebula to be associated with expanding material – by the American astronomer John Duncan in 1921; and the

FIG. 4.6   The Crab Nebula supernova remnant The Crab Nebula, designated as M1 and NGC 1952, is an expanding cloud of gas that was ejected by the supernova explosion detected by Chinese astronomers in 1054. Walter Baade took this visible-light image in 1955 using the 5.0-meter (200-inch) telescope on Mount Palomar, California. The filamentary gases are

expanding at a velocity of 1,500 kilometers per second. The red-yellow wisps of gas shine primarily in the light of hydrogen, while the blue-white light is the non-thermal radiation of high-speed electrons spiraling in magnetic fields. The south westernmost (*bottom right*) of the two central stars is the remnant neutron star of the supernova explosion, and a radio pulsar with a period of 0.33 seconds. This supernova remnant is located at a distance of about 6,300 light-years and has a diameter of 9.5 light-years. It is also a source of intense emission at radio, X-ray and gamma-ray wavelengths. (Courtesy of Hale Observatories.)

first to be recognized as the remnant of a stellar explosion – in 1928 by the American astronomer Edwin Hubble, who identified the Crab Nebula with the guest star recorded by Chinese astronomers in 1054 A.D. at the same region of the sky. The Crab supernova remnant is also the first radio source and the first X-ray source to be identified with a cosmic object outside the Solar System; the first remnant known to shine by non-thermal synchrotron radiation; and the first object known to be energized by a central pulsar.

The visible nebula consists of two distinct parts that emit radiation differently. A tangled, lacy network of red and green filaments, seen in the light of bright emission lines from ionized atoms, encases the inner, milk-white continuum radiation, which emits most of the energy from the Crab Nebula. This inner, visible light is impossible to reconcile with the nebula's intense radio emission through the radiation of a hot gas at any temperature. It is instead due to high-energy electrons spiraling around magnetic fields at nearly the velocity of light. This kind of emission is known as synchrotron radiation, since it was first observed in man-made synchrotron particle accelerators.

Unlike the thermal radiation of a hot gas, non-thermal synchrotron radiation increases in intensity at longer wavelengths. This is because more energetic, fast-moving electrons, which radiate at short wavelengths, lose their energy more rapidly than the slower moving electrons, which last longer and produce the nebula's long-wavelength radio emission.

At every wavelength of observation, the bulk of the radiation from the Crab Nebula is accounted for by the synchrotron radiation mechanism, including its X-rays, discovered in 1963. The very energetic electrons, which produce the exceptionally short-wavelength X-rays, should lose their energy and disappear in less than a year. After that, the Crab Nebula ought to radiate only at the longer wavelengths. But we know that the nebula keeps on emitting X-rays, so something must be replenishing the supply of energetic electrons on a continuous basis. Some enigmatic powerhouse located at the heart of the Crab Nebula had to be pumping energy into it, in the form of fast electrons. That source turned out to be a rotating neutron star, or pulsar, that was born in the crushed center of the explosion.

4.4 REBIRTH FROM THE ASHES – NEUTRON STARS AND PULSARS

Nothing is completely destroyed, and everything is transformed into something else, including stars. Having exhausted its nuclear fuel, at the endpoint of its bright, shining life, the former core of a massive star is crushed into oblivion by its own gravity. The old star's demise can nevertheless result in the simultaneous creation of a new star, like the resurrection of the magnificent, mythological bird, the phoenix, from the burning pyre of its ashes.

The death throes and final resting state of a star depends on its mass. Most stars are not massive enough to explode when they die. These Sun-like stars simply expand into giant stars, and when their time is up collapse into a burned-out, Earth-sized core known as a white dwarf. More massive giant stars, like Rigel and Betelgeuse, cannot stop at this stage when they run out of thermonuclear fuel. When the collapsing stellar core weighs between 1.4 and 3.0 solar masses, a neutron star is formed. It happens so quickly that the star's core shrinks to a city-sized neutron star in less than a second. When the collapsing stellar core weighs more than 3 solar masses, there is no end in site, and the core becomes a black hole.

Neutron Stars

Once protons were found within the nucleus of atoms, in 1920, scientists argued that there ought to be some other neutralizing particles down there to hold the nucleus together and keep the protons from pushing themselves apart. And after an eleven-year hunt, the English physicist James Chadwick discovered the neutron in 1932, by bombarding atoms with energetic particles, receiving the 1935 Nobel Prize in Physics for this accomplishment.

Just two years after the discovery of the neutron, Walter Baade and Fritz Zwicky suggested the possibility of a neutron star. They speculated that a supernova explosion is driven by the gravitational energy released when a normal star runs out of fuel and its core collapses, but that the explosion will not completely destroy the collapsing stellar core. A dense cinder could remain at the center, or in their prescient words: "With all reserve we advance the view that a super-nova represents the transition of an ordinary star into a *neutron* star, consisting mainly of neutrons. Such a star may possess a very small radius and an extremely high density."[45]

In a neutron star, gravity has overwhelmed the electrical force and crushed atoms out of existence, squeezing the electrons into protons until there is nothing left in the stellar core but neutrons. And the neutrons are compressed into a star about 10 kilometers in radius, with a mass density of about a billion billion, or 10^{18}, kilograms per cubic meter – a teaspoon-full of the neutron-star material would weigh about three billion tons on Earth.

The stability of such a dense state is related to the exclusion principle, first enunciated by the Austrian physicist Wolfgang Pauli in 1925. According to the Pauli exclusion principle, no two fundamental particles, such as electrons, can be in exactly the same place, or energy state, at precisely the same time. They instead resist being squeezed into each other's territory, darting away at high speeds just to keep their own space, somewhat like active dancers in a very crowded nightclub.

Of course, neutrons had not yet been discovered, or neutron stars imagined, at the time that Pauli enunciated his principle, but it nevertheless applies to them. Two neutrons cannot have identical energies, so they can't be pushed too close together, and as a result they produce an outward "degenerate" pressure that supports a neutron star.

Of course, you can't directly resolve an object 10 kilometers in size located out amongst the stars. So the very existence of these burnt-out, neutron-rich stellar cores remained conjectural until they were found by accident, as radio pulsars in 1967. But no one expected or predicted that neutron stars would emit the strong, distinctive radio pulses, and they were found using a telescope built for a completely different purpose.

Pulsar Emission from Isolated Neutron Stars

At a time when the clocklike radio pulses of pulsars were unknown and totally unforeseen, a group led by the English radio astronomer Anthony Hewish was studying the fluctuating intensity, or scintillations, of compact radio sources at 3.7-meters wavelength. The rapid and irregular fluctuations in the radio radiation are caused by motions of material in interplanetary space, driven by winds from the Sun. When the radio waves pass though the wind-driven material, they blink on and off, varying on time scales of a few tenths of a second, in much the same way that stars twinkle when seen through the Earth's varying atmosphere.

To study the radio scintillations produced by the solar wind, Hewish and his colleagues at the Mullard Radio Astronomy Observatory of Cambridge University constructed a gigantic array of 2,048 dipole antennas, each a simple wire on a pole, spread over four and a half acres, an area equivalent to fifty-seven tennis courts. The combined signals from all the antennas were connected to a detector whose output was rapidly recorded by wiggling marker pens on unwinding rolls of graph paper.

Repeated observations of several scintillating radio sources at different angles in relation to the Sun would tell something about the properties of the solar wind, while also providing information about the angular structure of the radio sources. The fluctuations are greatest for the smaller emitters, just as stars twinkle more than the Moon or planets, which have larger angular extents.

Susan Jocelyn Bell, from Belfast, Northern Ireland, would eventually play an important role in the discovery of pulsars. From reading her father's books, Jocelyn took an interest in astronomy at a young age, but she failed the examination required for students wanting to pursue higher education in British schools. So her family, whose Quaker heritage included strong support for women's education, sent her to a boarding school in England for a second chance.

After completing an undergraduate course in physics at the University of Glasgow in 1965, Jocelyn joined Hewish as a graduate student, becoming responsible for the network of cables connecting the dipoles in his scintillation instrument. And soon after the radio telescope was finished, in 1967, Jocelyn began to analyze the three-track records of graph paper by hand, examining about 30 meters of it every day, seven days a week for six months. Since a computer couldn't distinguish between the expected scintillations and man-made interference, she had to look and see what was happening, eliminating the terrestrial noise and detecting the varying radio signals as the rotating Earth pointed the giant radio telescope in the direction of different radio sources.

Within two months, she had found enigmatic "bits of scruff" that were fluctuating strongly in the middle of the night, when the Sun was not in the sky and the effects of its winds should have been small. Under the assumption that a flaring star emitted the signals, Bell and Hewish planned special observations to run the chart paper faster underneath the marking pens. This would spread the scruff out so you could see if it resembles well-known radio flares from the Sun. Each day for the better part of a month Jocelyn had to switch on the fast recorder when the source passed overhead, but it had turned off and disappeared from sight. Hewish was a bit peeved, telling Jocelyn "Oh, its a flare star; its gone and died and you have gone and missed it." But it came back to startle the world.

To everyone's amazement, the fast recorder revealed evenly spaced, periodic bursts of radio noise repeating at intervals of precisely 1.337279 seconds. Comparisons with precise terrestrial clocks showed that the periodic radio signal kept time with an accuracy of at least one part in one million. Only a very massive cosmic object could keep time with such accuracy. Moreover, each radio burst was brief and sharp, flashing on and off in milliseconds or less. Because nothing can move faster than the velocity of light, the varying source could not be bigger than the distance light travels in that time. So, the emitting source was not much larger than a planet, and might be much smaller.

In their discovery paper, published in 1968 and entitled "Observation of a Rapidly Pulsating Radio Source," Hewish, Bell and their colleagues named the object CP 1919, where CP stands for Cambridge Pulsar, and 1919 denotes one of its celestial coordinates, the right ascension of 19 hours 19 minutes. Today it is known as PSR 1919+21, with PSR designating pulsar, and the inclusion of the other sky coordinate, the declination of + 21 degrees.

The Cambridge radio astronomers also proposed a tentative explanation in terms of the pulsations of a white dwarf or neutron star. By this time they had found evidence for three similar objects that emitted rapid, periodic bursts of radio emission, like an extraordinarily precise cosmic clock. Astronomers were amazed, even incredulous. Fast variations with such regularity had never been observed before.

Large radio telescopes with sensitivity adequate for detecting pulsars had been in existence since the 1950s, but radio astronomers were used to adding up signals over long time intervals to smooth out random noise fluctuations and detect weak radiation. This precluded the detection of the pulsars, which are relatively weak radio sources when averaged over their periods. It is somewhat analogous to taking a long-exposure photograph that blurs shorter variations of commonplace events.

If time resolutions comparable to the pulsar burst durations of milliseconds had been used with existing large radio telescopes, the intense radio bursts would have been easily detected. To discover a radio pulsar you just need a big enough radio telescope to detect the faint signals, and a rapid recording device that will not smooth out the pulses. For this reason, many pulsars were quickly discovered by other radio astronomers in the months and years following the initial discovery by the Cambridge group.

It is actually because Hewish specifically designed a new type of radio telescope for a study of the rapidly-changing, solar wind effects that the pulsars were accidentally discovered. But he also had an innate curiosity and a prepared mind, which led Hewish and his colleagues to further investigate unexpected results that others might have ignored. In 1974 Hewish was awarded the Nobel Prize in Physics for his decisive role in the discovery of pulsars, but many felt that the Nobel Committee should have also recognized Jocelyn Bell.

We now know that the name pulsar is misleading, for the compact stars don't pulsate – they rotate, but the name has stuck. It designates repeating pulses of radio emission rather than a pulsating star.

In 1968, the same year as the announced discovery of pulsars, the Austrian-born American astronomer Thomas Gold proposed that the radio pulses are produced by a rotating neutron star, as a direct consequence of its rapid rotation and powerful magnetic field. He assumed that the pulsar would emit radio radiation in a beam, like a lighthouse, oriented along the magnetic axis (Fig. 4.7). An observer sees a pulse of radio radiation each time the rotating beam flicks across the Earth.

Since the time between successive pulses is the same as the rotation period of the neutron star, it had to be spinning around at a very fast rate, once every second or so. And because the neutron star's beam could be oriented at any angle, the beams of many pulsars would miss the Earth and a lot of them would therefore remain forever unseen.

Unlike many of today's astrophysicists and cosmologists, Gold suggested definitive observational tests of his ideas. He noticed that a spinning neutron star will gradually lose its rotational energy and slow down, successfully predicting that this would cause a slow lengthening of the radio pulsar periods with time. He also predicted that radio pulsars with much shorter periods would be found, as they were.

That doesn't mean that Gold was always right. He once incorrectly predicted that astronauts landing on the Moon would be buried under a thick layer of lunar dust. But at least he had ideas, and proposed ways to test them. He could even sleep through a fellow scientist's lecture, after a two-martini lunch, rising at the end of it to make an insightful comment.

If a slowly rotating star collapses down to a small size, the rate of rotation increases. An ice skater spins faster when pulling in her arms for the same reason. It's a result of the conservation of angular momentum, which means that the speed of rotation increases in direct proportion to the decrease in radius. So if an ordinary visible star runs out of nuclear fuel and is compressed to the size of a neutron star, its rotation period would speed up from days to seconds.

A neutron star is also a powerful magnet, and the magnetism is attached to the material within a star. When the star collapses it carries its magnetism with it, pack-

FIG. 4.7   Radio pulsar lighthouse A spinning neutron star has a powerful magnetic field whose axis intersects the north and south magnetic poles, and is offset from the spin axis that runs through the geographic poles. The rotating fields generate strong electric currents and accelerate electrons, which emit an intense, narrow beam of radio radiation from each magnetic pole. Because the magnetic field is misaligned with the star's spin axis, these beams wheel around the sky as the neutron star rotates, once per revolution. If one of the beams sweeps across the Earth, a bright pulse of radio emission, called a radio pulsar, is observed once per rotation of the neutron star. A wind of charged particles also streams from the poles but gets swept around to spray in all directions as the field rotates.

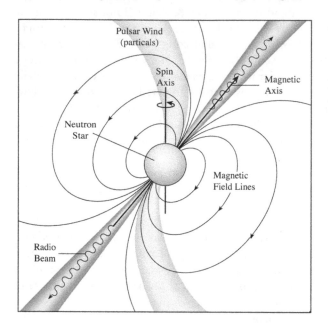

ing it into a smaller volume and amplifying its strength by factors of billions. The magnetic field strength increases in inverse proportion to the surface area, which is proportional to the square of the radius. So a typical magnetic field on a visible-light star, with a strength of 100 Gauss, would be strengthened to nearly a million million, or 10^{12}, Gauss if collapsing to a neutron star. This magnetism is at least a million times stronger than any magnetic field created in a terrestrial laboratory. And such a rapidly spinning and highly magnetized body is a powerful source of electromagnetic radiation.

The detection, in 1968, of a pulsar spinning 30 times a second in the Crab Nebula supernova remnant soon clinched the case for identifying pulsars with isolated, rotating, magnetized neutron stars. A white dwarf would be ripped apart if it rotated at this rate, but a smaller neutron star can spin thousands of times a second without destroying itself. Moreover, the period of the Crab Pulsar, designated PSR 0531+21, is increasing with time, just as Gold had predicted. Its period of 0.033 seconds is growing longer by 0.000 000 0364 seconds each day.

The Crab Nebula Pulsar is slowing down at a rate that indicates an age of roughly 1,000 years – close enough to the age of the supernova explosion observed by the Chinese in this region of the sky in 1054. Moreover, the loss of rotational energy inferred from the period increase of the pulsar is exactly that needed to keep the Crab Nebula shining, so it is the central pulsar that makes the nebula glow.

Nowadays, astronomers use the powerful *Chandra X-ray Observatory* to trace out the jets, rings, winds and shimmering shock waves generated by the highly magnetized, rapidly spinning Crab Pulsar (Fig. 4.8).

When the observed slow-down rate of most radio pulsars is combined with their periods, typical ages of between 1 million to 100 million years are obtained. A supernova remnant will expand and dissipate into the vastness of interstellar space, becoming unrecognizable in less than 100 thousand years, removing all signs of the pulsar's birth. Pulsars therefore outlive their supernova remnants and most pulsars are not found in one. Only a few rare, young pulsars, such as the Crab Pulsar with an estimated age of about 1,000 years and the Vela Pulsar, PSR 0833–45, with an age of roughly 10,000 years, are associated with visible supernova remnants.

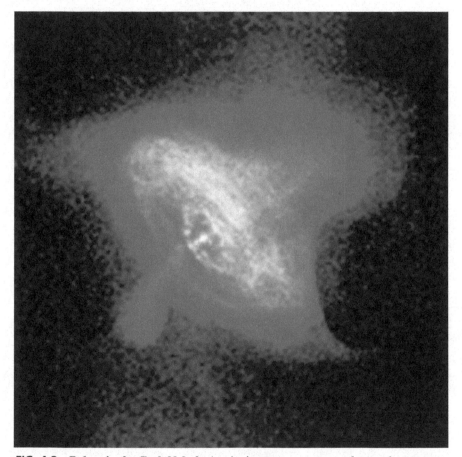

FIG. 4.8   Pulsar in the Crab Nebula A spinning neutron star, or pulsar, at the center of the Crab Nebula is accelerating particles up to the speed of light, and flinging them out into interstellar space. This X-ray image, taken from the *Chandra X-ray Observatory,* shows tilted rings or waves of high-energy particles that appear to have been flung outward from the central pulsar, as well as high-energy jets of particles in a direction perpendicular to the rings. The inner ring is about one light-year across. (Courtesy of NASA/CXO/SAO.)

Although most isolated neutron stars – those alone in space without a nearby companion – are only detected at radio wavelengths, the two associated with the Crab Nebula and Vela supernova remnants emit intense X-ray and gamma-ray pulses. They are rotating fast enough to efficiently accelerate particles to very high energies, radiating energetic X-rays and gamma rays.

Many other pulsars have now been detected at X-ray wavelengths, but unlike most radio pulsars, they are members of close binary-star systems rather than single isolated neutron stars. Before we turn our attention to these kinds of pulsars, we will first discuss a rapidly spinning radio pulsar with a nearby companion neutron star, which unexpectedly provided evidence for gravitational waves.

Gravitational Waves from PSR 1913+16

American radio astronomers Russell A. Hulse and Joseph A. Taylor discovered the now famous binary pulsar PSR 1913+16 in 1974, during a deliberate search for new pulsars using the latest computer technology with the huge radio antenna at Arecibo, Puerto Rico, the largest single radio telescope in the world. Their mini-computer was programmed to scan a range of possible pulsar periodicities, pulse durations and frequency dispersions, registering a signal whenever a pulsar passed through the telescope beam. After 14 months at Arecibo, and the discovery of 40 pulsars, Hulse, a graduate student at the University of Massachusetts at Amherst, found the enigmatic PSR 1913+16.

The pulsar's repetition period of 0.005898 seconds, or 58.98 milliseconds, had changed when Hulse next measured it, by just 0.00003 seconds. The clock-like signals of the other radio pulsars slowed down by much smaller amounts, so faulty detection equipment was at first suspected. But further measurements indicated an unsuspected regularity. The period change was itself cyclical, increasing and decreasing, rising and falling every 7 hours 45 minutes. This meant that the pulsar was in orbital motion, with the pulses being compressed together when the pulsar approached the Earth and pulled apart when moving away. As it turned out, PSR 1913+16 was in rapid orbit with another neutron star that did not emit detectable radio pulses, perhaps because its radiation beam is not aimed at the Earth.

Hulse completed his degree, and left the field of radio astronomy just a few years later. So precise timing of the radio pulses from PSR 1913+16 were continued by Hulse's advisor, Joe Taylor, and his other graduate students, permitting a determination of the orbital parameters of the system, as well as measurements of the mass of the pulsar and its silent companion. They weighed in at 1.44 and 1.39 times the mass of the Sun, as would be expected for two neutron stars.

More importantly, after four years of measurements and the analysis of about 5 million pulses, Taylor and his colleagues found that the orbital period was slowly becoming smaller, implying a more rapid orbital motion and a slow shrinking of the average orbit size. The two stars are drawing closer and closer, approaching each other at about a meter per year. This rate of orbital decay is just the change expected if their orbital energy is being radiated away in the form of gravitational waves, which had never been seen before.

Such gravitational waves were predicted in 1915 by Albert Einstein, who showed that any accelerating mass would emit them – as ripples in the curvature of space-time. Gravity waves travel at the speed of light, as electromagnetic radiation does. But while electromagnetic waves move through space, gravity waves are an agitation or undulation of space itself. They alternately squeeze and stretch the very fabric of space-time.

Gravity waves are produced whenever a mass moves, but they are exceedingly faint when generated and become diluted as they propagate into the increasing volume of space. They are so weak, and their interaction with matter so feeble, that Einstein himself questioned whether they would ever be detected. The gravitational radiation loss of the orbital energy of PSR 1913 + 16 nevertheless exactly matches the amount predicted by Einstein's theory, providing clear and strong evidence for the existence of gravitational radiation – for which Hulse and Taylor received the 1993 Nobel Prize in Physics.

But neither Einstein nor anyone else ever predicted that two neutron stars would be found that emit gravitational waves detected by timing the pulsar emission of one of them, and Taylor, Hulse and their colleagues did not set out to find a binary neutron star, let alone detect gravitational waves. It was another one of those serendipitous discoveries that make astronomy so wonderfully unexpected and surprising.

The discovery has stimulated attempts to directly measure gravitational radiation rather than just inferring its presence. The strongest gravity waves are expected to emanate from the most violent and abrupt cosmic events, when massive objects approach the speed of light. Intense gravity waves will, for example, be produced when a star explodes as a supernova or when neutron stars or black holes collide. Astronomers are now hoping to use the ambitious and costly Laser Interferometer Gravitational-wave Observatory, abbreviated LIGO, to detect the brief and slight stretching and compression that their gravity waves produce when arriving at Earth.

Reviving an Elderly Pulsar

About a thousand radio pulsars are now known, and almost all of them are isolated stars without a companion. As such a solitary pulsar ages, it gradually depletes its rotational energy and slows down, rotating with a longer and longer period. When entering old age, the pulsar eventually spins too slowly to induce the powerful voltage that sustains its radiation beams, and the neutron star fades into perpetual invisibility. This is apparently why there are no observed pulsars with periods longer than a few seconds or ages older than around 100 million years.

A single pulsar is then finished, no longer sending any messages to the outside world. But this does not have to be the pulsar's fate if a close stellar companion orbits it. Material pulled from this companion will revolve around the neutron star in the same direction that the star rotates, and spiral in to give the star a sideways blow that speeds it up. The in-falling matter can torque the elderly neutron star up until it spins a thousand times a second, even more rapidly than it did in its youth. The pulsar's dormant radio beams can then be resurrected from their early death. It's somewhat like an aging person being invigorated by a younger one, perhaps being rekindled by the fires of forgotten love.

Donald C. Backer of the University of California at Berkeley used the Arecibo radio telescope to discover such a radio pulsar in 1982, rotating with a period of precisely

0.001 557 806 448 873 seconds. When rounded off to just three significant figures, the period is 1.56 milliseconds, so it is known as a millisecond pulsar. And it's quite an amazing thing – a city-sized star with the mass of the Sun spinning around 667 times every second.

The millisecond pulsar would fly apart if it rotated at a slightly faster rate, so it is spinning just about as fast as it can. This means that it could not have been rotating much faster in the distant past, and could not have slowed down by much since then. Moreover, there is no supernova remnant anywhere near the millisecond pulsar, so it cannot be a young pulsar created during a relatively recent supernova explosion. So it's thought to be an elderly neutron star revived by matter falling onto it from a nearby stellar companion.

Binary star systems consisting of one neutron star in close orbit with an ordinary visible-light star also account for X-ray emitting pulsars.

X-rays from Neutron Stars

X-ray pulsars are amongst the brightest sources in the X-ray sky, and a famous example is Centaurus X-3, which pulses in X-rays every 4.84 seconds. Instruments aboard the first dedicated X-ray satellite, *Uhuru,* discovered the rapid pulsating variation shortly after the spacecraft's launch on 12 December 1970. After analyzing a year of observations of Centaurus X-3, the *Uhuru* scientists found a regular pattern of intensity changes of the X-ray pulses, increasing and decreasing in strength with a much longer period of 2.087 days and with systematic changes in the timing of the pulses at the same period. These effects were attributed to a companion star, which was orbiting the X-ray source and regularly eclipsing its emission.

Over the next year, accurate measurements showed that the average pulsation period of Centaurus X-3 was getting smaller, which meant that its rotation was speeding up and not slowing down like radio pulsars. So the rotational energy of the X-ray pulsar was increasing, rather than decreasing, with time.

The gain in rotational energy was interpreted in terms of matter drawn in from a nearby visible companion star. Since matter is being pulled toward the surface of an X-ray emitting neutron star, instead of being expelled from it, the neutron star is torqued up to a faster rotation. The material spirals in at the same direction as the neutron star's rotation, and when it lands it gives the neutron star a sideways kick, increasing its rotational energy and speeding it up.

As a result, the rotation periods of X-ray pulsars become shorter as time goes on. Because most radio pulsars are not members of binary systems, they expel material, lose rotational energy, slow down, and have periods that lengthen.

So a neutron star can be seen in X-rays when it is in a very close, orbital dance with a normal star that shines in visible light detected with an optical telescope. This stellar companion does not radiate detectable X-rays, but gas flowing from it fuels the X-ray emitting neutron star. This collapsed corpse, which has already entered the stellar graveyard, feeds on its more sedate companion, producing brilliant X-rays from the matter falling, or being pulled, into it.

And there are two ways of looking at the stellar duo. The visible picture portrays only the normal star, and the X-ray image just reveals its compact neighbor, the neutron star. A complete understanding of the double-star system, one visible and the other invisible, can only be understood when the two perspectives are combined.

Because the two stars are orbiting around each other rapidly, the gas does not fall directly onto the neutron star, but instead shoots past the neutron star, swinging around it like the arc of a welder's torch. The in-falling material forms a whirling disk of hot gas, known as an accretion disk, which spirals around and down onto the central neutron star, like soap suds swirling around a bathtub drain.

The inner portions of the swirling accretion disk revolve more rapidly than the outer portions, just as the closer planets orbit the Sun at faster speeds. The rapidly spinning inner parts of the disk constantly rub against the slower-moving outer parts. This viscous friction heats up the accretion disk and also causes the material in it to spiral inward. The closer the material moves toward the central neutron star, the hotter the in-falling gas becomes, eventually reaching temperatures of millions of kelvin and emitting luminous X-rays.

The intense magnetic field of a rotating neutron star also acts as a funnel to guide in-falling matter onto a neutron star's magnetic north and south poles, creating an X-ray pulsar (Fig. 4.9). As the accreting material heats up and falls onto the polar surfaces of the neutron star, it emits two beams of X-rays that flash in and out of view as the neutron star rotates and one or two of the beams sweep past the Earth. The powerful magnetic field of the neutron star also collimates the beams of a radio pulsar, but in this case the material is flowing out instead of in.

Sometimes a compact, invisible object with a visible companion can be force-fed with more matter than it can consume, and it hurls the in-falling matter out in two oppositely directed jets. This can happen when the invisible star is a black hole, which is the next topic in our discussion of cosmic violence.

4.5 BLACK HOLES – THE POINT OF NO RETURN

Imagining Black Holes

John Michell, a British clergyman and natural philosopher, suggested more than two centuries ago that certain stars could remain forever invisible. He reasoned that a star might be so massive, and its gravitational pull so powerful, that light could not escape it. As he wrote, in 1784, "all light from such a body would be made to return to it by its own proper gravity."[46]

The French astronomer and mathematician, Pierre Simon de Laplace popularized the idea, including it in his *Exposition du systeme du monde* in 1794, and subsequently showing that light could never move fast enough to escape the gravitational attraction of some compact stars. Their matter might be so concentrated, and the pull of gravity so great, that light could not emerge from them, making these stars forever dark and invisible.

Nowadays such a cosmic object is called a black hole, a term coined by the American physicist John Wheeler in 1967. It is black because no light can leave it, and it is a hole because nothing that falls into it can escape. So powerful is its gravity, that anything that draws near a black hole will be ripped apart and swallowed up, never to be seen again, at least in anything like the way it went in.

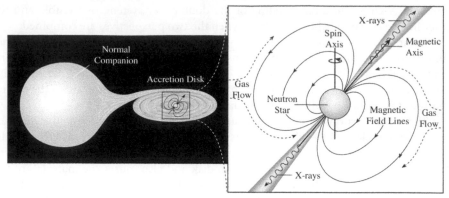

FIG. 4.9   X-Ray pulsar Material from a normal visible star spills over onto its companion, an invisible neutron star. The flow of gas is diverted by the powerful magnetic fields of the neutron star, which channels the infalling material onto the magnetic polar caps. The impact of the gas on the star's surface creates a pair of X-ray hotspots aligned along the magnetic axis at each magnetic cap. Because the magnetic field is not aligned with the neutron star's spin axis, the X-ray radiation from the hot spots sweeps across the sky once per revolution, producing the appearance of periodic X-ray pulsations if one of the hot spots intersects the observer's line of sight.

The name has nothing to do with the gloomy prison cell called the black hole of Calcutta, but it does have personal connotations and it has entered everyday language as a common metaphor. The black hole of loneliness is an example. A black hole might also remind people of their fears of losing something forever, like their childhood or their memory, or of being consumed, as in a job or marriage, or even destroyed, as in death. A black hole can also remind one of looking into the dark place at the center of someone's eye, which can look frighteningly empty.

Although Newton's theory of gravity was first used to show that black holes might exist, modern descriptions of black holes incorporate Albert Einstein's *General Theory of Relativity* in which mass and energy curve space-time. Nearby space is then said to curl into a black hole, carrying light and matter and any other form of energy with it. They are wrapped around a black hole so intensely that it becomes a cocoon shut off from the rest of the Universe.

While serving as an artillery lieutenant on the Russian front in 1916, the German astronomer and theoretician Karl Schwarzschild derived the solution to Einstein's equations outside a spherical, non-rotating black hole, showing that it contains a "singularity" at a radius that is now called the Schwarzschild radius in his honor. At this radius, the mathematics go berserk, time vanishes, compressed matter rips the fabric of space-time apart, and the known laws of physics break down – a singular place indeed.

Einstein disliked the notion that there might be a region of space that we could never learn anything about, and in 1939 he tried to prove that such bizarre singularities do not exist. But just a few month's after Einstein's attempted proof, the American physicist J. Robert Oppenheimer and his student Hartland S. Snyder used Einstein's own equations to demonstrate how such a black hole will form. In a paper entitled "On Continued Gravitational Contraction" they showed that a collapsing star could pass within the Schwarzschild radius, where the extreme warping of space would choke off the star's light, making it disappear as if swept up inside a magician's cloak.

Oppenheimer soon went off to direct the atomic bomb project at Los Alamos, New Mexico, and he never worked on black holes again. But scores of mathematicians and theoretical physicists pecked away at the problem, discovering all sorts of esoteric properties of black holes. They showed how the singularity at the Schwarzschild radius is a matter of perspective, using an appropriate coordinate system to place it at the center of the black hole down below the Schwarzschild radius.

You can't avoid the singularity completely, and you can't form a black hole without having one. It is the central place where all the mass is scrunched into a point of zero volume and infinite density, surrounded by empty space out to the hole's boundary, the event horizon, from within which nothing can escape.

The Schwarzschild radius marks the event horizon – quite literally a horizon in the geometry of space-time beyond which you can see no events, just as the Earth's horizon is the boundary for our vision. The curious thing about a terrestrial horizon is that its physical location becomes more distant when you approach it. In contrast, the event horizon of a black hole is a fixed place of no return. Everything inside it is disconnected from the outside, cut off forever from the rest of the Universe.

The event horizon resembles human death. Once you have crossed that line, you can't return. You are gone, lost, consigned to oblivion forever. For a star, only its gravitation remains, for humans it's memory by other persons and perhaps a soul.

The English theoretical astronomer Stephen W. Hawking has spent decades wondering about the fate of everything that has been swallowed up by a black hole. In 1974 he showed that very small black holes might evaporate or even explode, so they might not be so black after all, but his calculations showed that any radiation coming out of a black hole would contain no information about what went into it. Three decades later, in 2004, Hawking announced that all black holes will eventually deteriorate and send the matter and energy they consumed back into the place they came from, preserving the information that went in but in an unrecognizable state. So the complexity of the rest of the world, from insects to humans, and stars to galaxies, is still squeezed out of existence in a black hole.

Theoreticians have derived all kinds of theorems, laws and conjectures about black holes. In 1963, for example, the New Zealand mathematician Roy Kerr found analytic solutions to Einstein's equations of *General Relativity* that describe the curved space outside a rotating black hole. The event horizon is still there, but the rotation distorts it, causing it to bulge out at the equator with an oblate shape. Because of this rotation, any object that gets too close to a black hole is sucked into a whirling vortex, forced to swirl around the black hole as it falls into it. And this is related to something astronomers are more interested in, the actual observational evidence for the existence of black holes.

Observing Stellar Black Holes

How can we detect a black hole? With remarkable foresight, the British clergyman John Michell also speculated, in 1784, that the unseen star might betray its presence by its gravitational effects on a nearby, luminous star in orbit around it. In modern extensions of this idea, a black hole uses its powerful gravity to suck in the outer layers of a nearby visible star. The in-falling material swirls around and down into the black hole, like new snow falling into the dark of night, orbiting faster and faster as it gets closer – as the result of the ever-increasing gravitational forces. The rapidly moving particles collide with each other as they are compressed to fit into the hole, heating the material to millions of kelvin. At these temperatures, the gas emits almost all of its radiation at X-ray wavelengths. So, the way to find a black hole is to look for a perfectly ordinary star that orbits an invisible nothing that also emits powerful X-rays.

The heated material does not fall directly into the black hole; it swirls around it. So you can picture the central black hole as a formless, engulfing void, surrounded by an X-ray vortex, whirling at the edge of oblivion. These X-rays are emitted as a final gurgle, a sort of last hurrah, just before the swirling material disappears into the black hole, resembling the final brilliant flash of a light bulb just before it burns out.

The archetype of a stellar black hole is Cygnus X-1, located in the constellation Cygnus and one of the first X-ray sources to be discovered. Rapid, irregular X-ray bursts from Cygnus X-1, detected from the *Uhuru* satellite in 1970, varied so quickly that it had to be an extremely compact stellar object. The X-rays flickered on and off as rapidly as a few milliseconds, and because nothing travels faster than the velocity of light the emitter had to be less than 300 kilometers across. This meant that it was smaller than a white dwarf star, and had to be either a neutron star or a black hole.

Observations of the orbital characteristics of the visible companion of Cygnus X-1, in 1972 by the English astronomers B. Louise Webster and Paul Murdin and independently confirmed by the Canadian astronomer Charles Thomas Bolton, provided a firm lower limit to the mass of the X-ray source. It had to be more than 8 times the mass of the Sun. Any normal star with this mass would be very bright and easily seen through a telescope, but it is dark, emitting no detectable visible light, and above the neutron star mass limit of 3 solar masses. It had to be more compact than a white dwarf, and its mass is greater than that of a neutron star, so by elimination it must be a black hole.

We now know of at least ten stellar black holes, identified in this way. The orbital properties of visible companions indicate masses between 4 and 12 times that of the Sun, beyond the neutron star limit. Like Cygnus X-1, many of these black holes also exhibit highly luminous, rapid and irregular X-ray outbursts, showing that they are very small on a cosmic scale. The transient X-ray flickering, sometimes brightening a million-fold in milliseconds, is most likely emitted as in-falling material takes the final plunge and vanishes through the event horizon of a black hole.

Super-Massive Black Holes

Exceptionally formidable black holes are found at the centers of nearby galaxies. They are very massive, scaled-up versions of stellar black holes with millions if not billions of times the mass of the Sun, packed into a region just a few light-years across. Like their stellar counterparts, these super-massive black holes cannot be directly observed. Their presence is indirectly inferred from whirlpools of stars and gas whose velocities rise sharply toward the center of the galaxy, to values that indicate an unseen super-massive black hole is hidden there. Without its gravitational pull, the surrounding material would fly away from the galaxy, like a high-speed rocket sent into outer space.

The most compelling and earliest evidence for a dark, super-massive object came from observations of the core of M87, a large elliptical galaxy located some 50 million light-years away from us, sitting at the center of the giant Virgo cluster of galaxies. M87 is also a very strong radio source, called Virgo A and catalogued as 3C 274, and has unusual jets of matter that emanate from its core.

Visible-light spectral observations, carried out in the late 1970s by the English-born American astronomer Wallace Sargent and his collaborators, indicated that the velocities of stars orbiting the center of M87 increase strongly toward the center. Billions of solar masses had to be crammed into a very small volume, no bigger than 300 light-years across, to keep the high-velocity stars from flying off into intergalactic space.

Confirming evidence of a colossal black hole at the core of M87 was provided in 1994, when instruments aboard the *Hubble Space Telescope* found a glowing, rotating disk of gas near the center of the galaxy. Only the gravitational pull of a central, exceedingly massive, black hole, weighing 2.4 billion solar masses, can hold the whirling gas together.

So far, observational evidence for super-massive black holes has only been obtained in our local cosmic surroundings. This is because only relatively nearby galaxies are close enough to view their central regions with the detail needed to detect material whirling about their centers. But telescopes are also time machines, detecting radiation that was generated long ago in the youth of more distant objects, sometimes billions of years in the past. And the observations indicate that the younger, more remote objects are much more energetic than the older, nearby ones. The most persuasive way to explain the tremendous energies and hot jets spewing from these active objects is a super-massive black hole.

4.6 COSMIC RAGE – RADIO GALAXIES, QUASARS AND GAMMA-RAY BURSTS

Radio Galaxies – Active Cores, Dual Jets and Twin Radio Lobes

During recent decades, astronomers have discovered a new type of cosmic violence, the powerful, remote radio galaxies. They emit a million times as much power at radio wavelengths as the Milky Way and other nearby spiral galaxies, and are often associated with elliptical galaxies detected using optical telescopes at visible-light wavelengths. When the optical image is combined with maps of the radio signals, it is found that the radio emission is not confined to its visible counterpart, but instead concentrated in two radio lobes that are separated from the central visible galaxy by hundreds of thousands of light-years. It is as if the radio-emitting clouds were expelled from the central galaxy, which is detectable only at visible wavelengths.

The American astronomer Carl K. Seyfert published early evidence for violent upheaval in some galaxies in 1943. He showed that a few percent of spiral galaxies, subsequently named after him, have small, intense, blue centers that radiate spectral lines of the type produced by ionized gas. Instead of being narrow, the motions of the high-speed ions widened these lines, and the velocities implied from their widths, when interpreted by the Doppler effect, are several percent of the speed of light, sometimes well in excess of that required to escape the gravitational bonds of the entire galaxy. Therefore, their matter must be flowing out into intergalactic space, and some of the Seyfert galaxies do exhibit bright filaments that suggest the ejection of gas (Fig. 4.10). Full-fledged radio galaxies, with radio jets and lobes fed by high-speed electrons, were subsequently discovered surrounding and fed by Seyfert galaxies.

The prototype for the more powerful, double-lobed radio galaxies is Cygnus A (Fig. 4.11), listed as 3C 405, the 405th source in the third Cambridge, or 3C, catalogue. Its astonishing radio power, equivalent to the visible luminosity of a million, million stars, is attributed to the non-thermal synchrotron radiation of high-speed electrons supplied from the visible center along two oppositely directed jets that feed the radio lobes. These dual jets remain extraordinarily straight and surprisingly stable across distances as large as a million light-years, energizing the radio lobes and pushing them further and further apart.

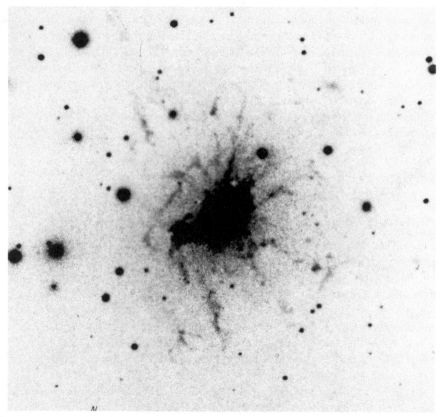

FIG. 4.10 **Seyfert galaxy** A negative print of the visible hydrogen emission from the Seyfert galaxy NGC 1275. (Courtesy of Kitt Peak National Observatory.)

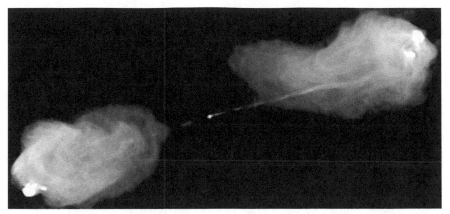

FIG. 4.11   Radio galaxy Cygnus A This radio image, taken with the Very Large Array, reveals the double-lobed radio structure of the radio galaxy Cygnus A, listed as 3C 405 in the third Cambridge survey and the strongest extraterrestrial radio source in the sky. It lies at a distance of about 750 million light-years and has a radio output a million times more powerful than the radio emission of a normal galaxy like the Milky Way. The radio galaxy extends about half a million light-years from end to end. Narrow, straight radio-emitting jets of fast-moving particles have probably been ejected along the rotation axis of a supermassive black hole located within a central elliptical galaxy. It must have been active for tens of millions of years to produce the two radio lobes. The jets bore through the thin gases in intergalactic space until they pile up, terminating in "hot spots" and back-flowing to fill up the lobe regions with wispy cocoons of enhanced emission. (Courtesy of NRAO/AUI/NSF.)

If the radio galaxy has been sending out radio power at the present rate over its estimated million-year lifetime, it has emitted an enormous total radio energy equivalent to the complete annihilation of about a hundred thousand stars. This energy is so colossal that relatively inefficient nuclear processes, which make stars shine, seem to be ruled out. Moreover, all this energy is somehow being channeled primarily into the form of particles moving very close to the velocity of light along well-defined, narrow jets. They are most likely anchored in a gigantic, spinning black hole, which acts as a gyroscope and ejects high-speed particles into the jets.

Radio arrays of increasing sophistication and sensitivity, such as the Very Large Array, abbreviated VLA, have provided detailed information about the lobes, jets and central cores of the double-lobed radio galaxies, which exhibit a variety of shapes depending on the jets' internal strength and external forces, as well as our viewing perspective. And as the VLA was probing the detailed structures of radio galaxies, even more dramatic sources of energy were found still deeper in space, located in the cores of quasi-stellar sources, or quasars for short.

Quasars

The discovery of quasars was a consequence of an ongoing search for the visible-light counterparts of powerful radio sources in the early 1960s, in which radio interferometers were used to provide accurate locations that might permit optical identifications in the sky seen in visible light. The young radio astronomer Thomas A. Matthews, at the California Institute of Technology, used an interferometer to refine the radio positions of discrete radio sources with small angular sizes, and presumably great distances, including 3C 48 – the 48th source in the third Cambridge catalogue, which then had no measurable radio extent at all. Armed with the location, Allan R. Sandage pointed the 5.0-meter (200-inch) Palomar telescope in the direction of 3C 48, but there was no galaxy to be found. Instead, Sandage discovered a bright blue object, no bigger in angular size than a star, and with a totally confusing spectrum that was unlike anything ever seen before.

Subsequent monitoring by Sandage indicated that its visible brightness varied as much as a factor of two from one fortnight to the next. Since the light from some stars varies on similar time scales, and galaxies do not, the strange object seemed to be a star. Sandage and his collaborators therefore reported in 1960 that 3C 48 might be the first true "radio star" – other than the Sun. But they were so mystified by the nature of the object, that they never published the customary abstract of their paper; the results remained unpublished in the professional journals for two years, and the true nature of 3C 48 remained unknown even after the formal paper announcing these results was written.

The key to the mystery was provided when the Moon happened to pass in front of another bright radio source, 3C 273, in 1962. The radio astronomer Cyril Hazard, then at the University of Sydney in Australia, realized that a careful timing of the disappearance and reappearance of the occulted source would establish a precise position, since the location of the Moon's edge is known accurately for any time.

So Hazard and his colleagues used the occultation method to show that 3C 273 is a double radio source, one component of which apparently coincides with a blue stellar object. This coincidence prompted Maarten Schmidt to confirm the identification and obtain an optical spectrum using the 200-inch telescope. The spectra indicated a completely unexpected and exceptionally high recession velocity of 0.16 percent of the velocity of light. When he told his colleague Jesse Greenstein about the discovery, Greenstein produced a list of emission line wavelengths for 3C 48, and within minutes they had found that it is rushing away with an even faster motion at 37 percent of the velocity of light.

When these velocities are used to infer distances using Hubble's law, it is found that 3C 48 and 3C 273 are located at distances of billions of light-years. And when their observed luminosities are combined with these distances, the intrinsic visible-light luminosities are comparable to that of 10 million million, or 10^{13}, Sun-like stars.

Since the bright objects appeared star-like in visible light, they became known as quasi-stellar radio sources, a term that was soon shortened to quasars. The quasars had, in fact, been ignored as stars on optical photographs for years. But a casual inspection of the optical sky would never have led to the discovery of quasars. About 3 million stars look brighter than the brightest quasar, 3C 273.

In contrast, the powerful observed radio emission of 3C 273 makes it easy to locate at radio wavelengths; only a handful of such sources shine so brightly in the radio sky. That's why radio astronomers were destined to point the way to the first quasars. If radio astronomy had developed later, X-ray astronomy would have played the same role for the same reason. Unlike an optical photograph of the sky, which is dominated by the visible light of ordinary stars, an X-ray image can consist mainly of quasars, which can have either strong or weak radio emission.

Once quasars were known, astronomers located others by obtaining optical spectra of blue, star-like objects that are located well outside the plane of the Milky Way, where stars are not supposed to be, and measuring the large redshifts characteristic of remote quasars. Thousands of quasars have now been discovered in this way, some of them emitting intense radio signals and other silent ones with their radios turned off.

Quasars become increasingly numerous as we look deeper into space. They are located at the very fringe of the observable Universe, beyond many radio galaxies and over on the other side of the Universe (Fig. 4.12). So quasars appear as they were about 10 billion years ago, when the radiation now reaching the Earth began its journey, but they are seldom found at closer distances corresponding to an older age. The youthful spurt of quasar activity gets worn out and used up as galaxies grow older.

Astronomers have gradually come to realize that quasars are brilliant, tiny cores embedded in much larger, extremely active galaxies, whose outer parts are difficult to detect in the intense quasar glare. The reason that quasars appear to us as small objects is that they greatly outshine the surrounding galaxy, rendering it invisible. Looking at them is like driving directly into the bright headlights of an oncoming car, which blind us from seeing anything else. From its vantage point in space, the *Hubble Space Telescope* has resolved the core quasar light, and removed it from the computerized images to detect the faint, fuzzy halo of a host galaxy. Its overall visible extent is as large as the

FIG. 4.12 Composite Hubble diagram Because of the expanding Universe, redshift and distance are intimately related. The farther away an object is, the more its light has been stretched, or redshifted. Here the distance is expressed in terms of the look-back time, the time that light took to travel from an object to the Earth. Quasars have the largest redshifts, and the light we now see from the most distant quasars was emitted when the Universe was less than half its present age. Radio

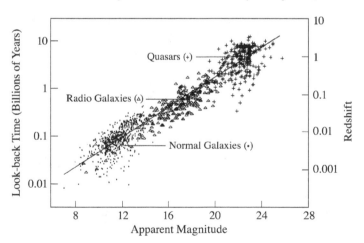

galaxies also exhibit large redshifts, but they are generally smaller than those of quasars. Normal galaxies, with relatively weak radio emission, usually have redshifts less than one, which means that they are relatively nearby and that the light we now see from them was emitted relatively recently, from objects that were then much older than either quasars or radio galaxies.

elliptical galaxies found at the centers of many radio galaxies. More extensive radio lobes have also been found for some quasars, connected by radio emitting jets to their compact, central source.

Black-Hole Power

What energizes a quasar, which can be as small as the Solar System and shine with the light of 10 million million stars? The astrophysicists Edwin E. Salpeter, at Cornell University, and Yakov B. Zeldovich in Moscow, independently proposed in 1964 that the fantastic quasar energy is due to an extremely massive black hole, which emits radiant energy while consuming nearby gas and stars. The gravitational pull of a mass equivalent to 100 million Suns is needed to balance the visible quasar luminosity; otherwise its radiation pressure would just blow the quasar away. Such a black hole would be sufficiently small and powerful enough to explain the tiny sizes and the colossal brightness of quasars.

The tremendous luminosity of every radio galaxy and quasar is most likely supplied by a compact, super-massive black hole. Its powerful gravity pulls in surrounding stars and gas, forming a flat, orbiting accretion disk that spirals into the black hole. As proposed by Roger D. Blandford and Martin J. Rees in 1974, with subsequent contributions by Mitchell C. Begelman, Roman Znajek and others, powerful magnetic fields generated by the rotating black hole turn the whirling accretion disk into an enormous dynamo. It uses the hole's rotational energy to accelerate charged particles and squirt them out in diametrically opposite directions along the rotation axis at about the speed of light. They continuously feed the two radio lobes commonly found symmetrically placed from the center of radio galaxies and quasars.

The classic example is M87, a giant elliptical galaxy whose central spinning disk of hot gas indicates that a super-massive black hole resides at its center. A one-sided jet of gas emerges nearly perpendicular to the disk, and stretches out into one of the two lobes of the radio galaxy Virgo A, numbered 3C 274 in the Cambridge survey (Fig. 4.13). The motions of bright knots in the jet indicate that they are traveling outwards at about half the speed of light. And Very Long Baseline Interferometry observations with widely separated radio telescopes reveal that M87's jet emerges from a region at most six light-years across, most likely harboring the super-massive black hole that produces the jet.

To power the youthful activity of a quasar by accretion, there has to be about one solar mass per year of gas flowing into the black hole. So billions of stars or the equivalent amount of gas must be consumed as a quasar or radio galaxy evolves over the course of billions of years. The supply dwindles away over time and the activity dies down, but the black hole does not disappear. Most galaxies probably contain super-massive black holes at their center. The ones in the older, nearby galaxies are the starving remains of former quasars, with a dwindling supply of material that once fed a higher rate of activity. They are found in ordinary nearby galaxies, like Andromeda and the Milky Way, whose cores are old surviving fossils of former quasars.

Yet, there are other ways to attain comparable energy over very brief intervals of time, lasting seconds or minutes rather than billions of years. This is in the violent explosions that create the gamma-ray bursts.

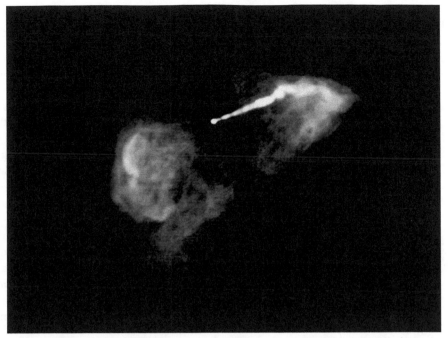

FIG. 4.13   Radio jet from M87 The bright radio source Virgo A, also designated 3C 274, coincides with M87, a giant elliptical galaxy located at a distance of about 50 million light-years in the Virgo cluster of galaxies. This radio map, made with the Very Large Array, shows two elongated lobes, one on either side of the center of M87. The most intense radio emission comes from a jet that emerges from the core of the galaxy and stretches some 8,000 light-years into one of the two lobes. The observed high-speed motion of bright knots in the jet implies that its radio-emitting electrons are traveling at nearly the speed of light. The visible light emitted by the jet was first noticed in 1917, by Heber Curtis at the Lick Observatory in California, who wrote that: "a curious straight ray lies in a gap in the nebulosity ... apparently connected with the nucleus by a thin line of matter. The ray is brightest at its inner end."[b] Observations of the rates at which stars and gas clouds revolve within the central core of M87 indicate that it contains a compact massive object, most probably a supermassive black hole of about 3 billion solar masses. (Courtesy of NRAO/AUI/NSF.)

Gamma-Ray Bursts

At least once a day, an explosion somewhere in the depths of outer space produces a brief, intense blast of high-energy, gamma-ray radiation, but never from the same place twice. They are the most powerful kind of cosmic violence known so far, surpassed only by the Big Bang that set the Universe in motion about 14 billion years ago.

These spectacular gamma-ray bursts were discovered by accident in the late 1960s, using the *Vela* series of satellites belonging to the Department of Defense under the auspices of the United States Air Force. They were monitoring Soviet compliance with an international treaty banning the tests of nuclear warheads in either the Earth's atmosphere

[b] Heber Doust Curtis (1872–1942). "Descriptions of 762 nebulae and clusters photographed with the Crossley reflector," *Publications of the Lick Observatory* **13**, 31 (1917)

or outer space, hence the name Vela from the Spanish word *velar,* which means "to guard." The Russians might, for example, carry out nuclear tests on the far side of the Moon, which cannot be seen from the Earth, but the *Vela* satellites could detect gamma rays emitted from the cloud of radioactive material blown out beyond the Moon's edge after the nuclear blast. So the satellites had instruments designed to look for gamma rays, coming from any direction and indicating a clandestine nuclear test.

The secret satellites were launched and operated in identical pairs on opposite sides of a circular orbit, eventually with sufficient time resolution to determine the direction of the source from the difference in arrival time of its radiation at two or more satellites. The sensitive detectors on the last two double satellites, the *Vela 5* and 6 pairs launched on 23 May 1969 and 8 April 1970, monitored incoming gamma rays every one sixty-fourth of a second, which was rapid enough to discriminate between the direction of a possible enemy nuclear explosion and cosmic sources of gamma rays such as solar flares and supernova explosions.

The *Vela* data were analyzed at the Los Alamos National Laboratory in New Mexico, where the world's first atomic bombs had been developed during World War II. Although a clandestine nuclear explosion was never detected, unexpected flashes of gamma rays were discovered coming from different places in deep space, with an existence measured in seconds. Moreover, the lag between the burst arrival times at the two defense satellites indicated that they originated far beyond the Solar System, and were therefore of cosmic origin.

The gamma-ray bursts were nevertheless kept secret for about five years, until the Los Alamos scientists Ray W. Klebesadel, Ian B. Strong and Roy A. Olson described the discovery on 1 June 1973, in the *Astrophysical Journal Letters.* Civilian astronomers might blame the military for keeping such an important discovery hidden for years, but it is also likely that the gamma-ray bursts would have never been discovered if it wasn't for the defense satellites. No other government agency was likely to fund a speculative program designed to look for such totally unknown, unsuspected, and unprecedented events in outer space.

The nature of the enigmatic gamma-ray bursts remained a mystery for more than two decades after their discovery. During all that time, no counterpart was ever observed at any other wavelength, in X-rays, visible light or radio waves. So no one knew how far away they were, or had any substantive clue about the mechanism that might produce them. They just flashed on for a few milliseconds, seconds or minutes, briefly outshining all other sources in the gamma-ray sky combined, and then disappeared forever, never to appear again at that point in the sky – like the last shouts of someone disappearing into the void.

Owing to the lack of observational data to test their ideas, speculative theorists went wild. They published hundreds of explanations for gamma-ray bursts in the 1970s and 1980s. Most of these imaginative conjectures included two assumptions. The unknown sources were thought to be relatively nearby, originating within our Galaxy, for excessive output energies would be required if they were further away. And because the gamma-ray bursts flashed on in less than a millisecond, or less than 0.01 second, and sometimes lasted for just a few milliseconds, everyone assumed that they originated in relatively compact objects. But other than their supposed proximity and probable small size, there was no consensus about the mechanism that powered the brief, energetic outbursts, just one at a time. A comet or asteroid colliding with a neutron star, quakes in a neutron star's interior, or a flaring explosion in the intense magnetic fields of a neutron star were amongst the theories.

Further observations, which might constrain theoretical conjectures, were not made until the launch of NASA's *Compton Gamma Ray Observatory,* abbreviated *CGRO,* on 5 April 1991 from the *Space Shuttle Atlantis.* Its all-sky detectors, placed at each corner of the box-like spacecraft, were about one hundred times more sensitive than the *Vela* ones, so they could detect fainter gamma-ray bursts that occur more frequently. *CGRO* detected a new gamma-ray burst almost every day, compared with once a month for the *Vela* instruments. Moreover, by comparing the strength of the signals in the different detectors, a rough direction could be determined, accurate to a few degrees.

In less than half a year after the launch of the *CGRO,* NASA scientists Charles A. Meegan and Gerald J. Fishman announced that the 141 gamma-ray bursts detected so far were scattered randomly over the sky, with any general direction having roughly the same number. This conclusion remained unchanged for the 2,704 bursts observed by the spacecraft during its nine-year lifetime (Fig. 4.14). There is no concentration of the burst sources in that sky-girdling band of stars known as the Milky Way, or at the center of our Galaxy, or anywhere else.

North Galactic Pole

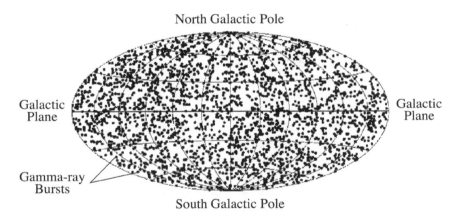

Galactic Plane

Galactic Plane

Gamma-ray Bursts

South Galactic Pole

FIG. 4.14 Gamma-ray burst distribution The positions of 2,704 gamma-ray bursts observed from the *Compton Gamma Ray Observatory* between 1991 and 2000, plotted in galactic coordinates. Each dot corresponds to a single gamma-ray burst. The distribution is consistent with an isotropic, or random, distribution over the sky, with no statistically significant clustering in a preferred direction. In particular, there is no preferential concentration of bursts along the plane of our Galaxy, which runs horizontally through the middle of the diagram. This suggests that the gamma-ray bursts do not belong to our Galaxy, since they are not concentrated in the Milky Way plane where most stars are found. These data were taken with the Burst And Transient SourcE, abbreviated BATSE, instrument aboard the spacecraft. (Courtesy of NASA, *CGRO,* and the BATSE team.)

Such an isotropic distribution had been used in previous "great debates" to suggest that spiral nebulae are "island universes", or galaxies, that lie outside the Milky Way, and that powerful radio sources are not stars but instead distant galaxies.

Nevertheless, because they could not be associated with any other cosmic object, astronomers could not definitely prove that the gamma-ray bursts are located so far away. Scores of possible visible-light counterparts could be found in the large, degree-wide patches of sky seen by the instruments aboard the *Compton Gamma Ray Observatory*. Efforts to establish precisely where the mysterious bursts came from were also frustrated by the unpredictable, rapid, transient and non-repeating nature of the bursts. Since they appeared without warning, disappeared from view in seconds or minutes, and never returned, it was hard to obtain definitive information about them.

Then the Italian-Dutch *BeppoSAX* satellite was used to pinpoint a more exact location of the fleeting bursts. Its peculiar acronym combines the nickname "Beppo" of the Italian physicist Giuseppe Occhialini with *SAX*, an abbreviation of *Satellite per Astronomia X*. After a ten-year delay and soaring costs, with a huge budget overrun, *BeppoSAX* was finally launched on 30 April 1996 from the Kennedy Space Center in Florida. It enabled astronomers to solve most of the fundamental mysteries about gamma-ray bursts during the following year, when it was conclusively shown that they are emitted from enormously distant galaxies.

As soon as instruments on *BeppoSAX* picked up a gamma-ray burst, instruments on the satellite also detected its X-ray emission, which can last for hours, and fixed its location with a precision of a few minutes of arc. With these accurate coordinates, made available by telephone or e-mail, other astronomers were able to point powerful ground-based telescopes at the same spot in the sky, detecting faint, long-lived emission at visible-light and radio wavelengths.

It wasn't until ten months after launch that the relatively accurate location of one gamma-ray burst, occurring in 1997 on February 28 and designated GRB 970228 for the date, was used to detect the first visible-light counterpart of a gamma-ray burst, as a fuzzy patch of light emitted by an expanding, cooling fireball that can last for days or even weeks, like the faint glow of the dying embers of a fire. Within a year, astronomers had measured the redshifts of the host galaxies of several gamma-ray bursts, including the colossal redshift of 3.42 obtained for GRB 971214. And that measurement wasn't easy; it was something like detecting a 100-watt light bulb at a distance of nearly a million kilometers. The large redshift indicated that the galaxy that spawned this gamma-ray burst is located at a staggering distance of 12 billion light-years, moving at almost the velocity of light near the apparent edge of the observable Universe.

To be seen from so far away, GRB 971214 had to be incredibly powerful. For a few seconds, the total gamma-ray energy emitted by the burst was comparable to the visible output of 100 billion billion, or 10^{20}, stars, or about as much light as that emitted by all the stars in all the galaxies in the observable Universe.

How can so much energy be released in so short of time? It might be caused by the catastrophic death of a massive star, or it could result from the collision of two neutron stars. Both mechanisms could be suddenly switched on in a millisecond or less, they could each fade away in an expanding fireball, creating weak afterglows, and both of them might be associated with the birth of a black hole.

When the core of an exceptionally massive star runs out of time and collapses into a rotating black hole, its outer layers could be expelled in a supernova explosion while the collapsing material swirls down and around the central black hole and is hurled back out in jets along its rotation axis. A short-lived burst of gamma rays might be emitted when the jets bore through the outer layers of the star and plow into surrounding interstellar material at nearly the speed of light (Fig. 4.15), also producing a fireball that fades as an afterglow and eventually reveals the expanding debris of the exploded star.

No afterglows of the shortest gamma-ray bursts, lasting less than two seconds, have ever been observed, and the merging neutron star hypothesis remains a viable alternative for them. The two stars would spiral together, slowly at first and then faster and faster, until they finally merged. As the two stars meet, their enormous gravity will tear them apart, most likely disappearing into a black hole accompanied by twin jets that account for the gamma-ray bursts.

And this brings us to the most energetic explosion of them all, the Big Bang, which occurred about 14 billion years ago, setting the Universe in motion.

4.7 THE BIG BANG

The Beginning of Time

Regardless of the direction in which we look out into space, all the distant galaxies are flying apart, dispersing and running away from us at speeds that increase with their distance, as if they had been ejected from a common point at some time in the past. This is characteristic of an explosion – the farther apart one particle of debris is from another, the faster they move from each other. It is as if the observable Universe

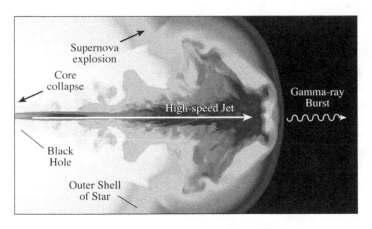

FIG. 4.15 Gamma-ray burst A computer simulation of a high-speed jet formed after the collapse of a massive star and associated supernova explosion. The core of the star has collapsed to form a black hole at the center, surrounded by a disk of accreting matter. A jet of matter is launched away from the black hole, breaking through the outer shell of the star about nine seconds after the jet's creation. It is moving to the right, in the direction that coincides with the rotational axis of the star. When moving at nearly the speed of light, the jet will make a gamma-ray burst. The dense plug at the head of the jet and the cocoon around the jet beam are expected to emit X-rays. (Courtesy of Weiqun Zhang and Stan E. Woosley, the University of California at Santa Cruz.)

resulted from the detonation of a cosmic bomb, the Big Bang, and we are participating in the explosion, watching it blow up before our very eyes.

The American poet Robinson Jeffers, whose brother was an astronomer at Lick Observatory, has captured its essence with:

> Nothing can hold them down; there is no way to express that explosion;
> all that exists
> Roars into flame, the tortured fragments rush away from
> each other into all the sky, new universes
> Jewel the black breast of night; and far off the outer nebulae
> like charging spearmen again
> Invade emptiness.[47]

Another beautiful description of the explosive beginning, and subsequent expansion, of the observable Universe was written by the Belgian astrophysicist and Catholic priest Georges Lemaître in 1931, shortly after the discovery of the expanding Universe: "The evolution of the world can be compared to a display of fireworks that has just ended: some few red wisps, ashes and smoke. Standing on a well-chilled cinder, we see the slow fading of the suns, and we try to recall the vanished brilliance of the origin of the worlds."[48]

As Lemaître realized, it was radiation that dominated the expansion in its earliest stages, perhaps even before matter was created. So just after the Big Bang, radiation was the most powerful thing around. Pure and incredibly energetic radiation was even being transformed into matter! That radiation was so intense that it is still around, and we're immersed within it.

Three Degrees Above Zero

Thirty years after Karl Jansky's serendipitous discovery of cosmic radio static, Arno A. Penzias and Robert W. Wilson accidentally discovered the three-degree cosmic microwave background radiation, or just the background radiation for short. It is the cooled relic heat of the Big Bang that gave rise to the expanding Universe. Like Jansky, Penzias and Wilson were working at the Bell Telephone Laboratories, abbreviated Bell Labs, trying to track down noise that was interfering with radio signals, but at a much shorter, microwave wavelength of 7.35 centimeters.

The discovery of this cosmic microwave background radiation was an unexpected consequence of technology developed for military radar and the civilian communication industry. During World War II, when Bell Labs was heavily involved in developing radar at centimeter wavelengths, their radio engineers invented a new kind of sensitive antenna to detect weak signals, known as a horn-reflector antenna because its shape resembles that of an old-fashioned ear horn. Two decades later, when the laboratory became interested in communication satellites, they built a larger horn antenna to receive faint return signals when bouncing microwaves off a plastic balloon, known as *Echo*.

The sensitive antenna was very effective in rejecting radiation from the surrounding ground, whose temperature is about 300 kelvin, picking up less than 0.3 kelvin. An ordinary radio antenna might receive 30 kelvin from the ground, creating more confusing noise and making it harder to detect a faint signal. A unique low-temperature, maser receiver was also attached to the giant horn, and used to amplify the weak signals reflected by *Echo*.

It worked just fine for the *Echo* experiment, and was subsequently used, in 1962, to receive a signal sent down from the *Telstar* satellite at a wavelength of 7.35 centimeters. And the successful tests of *Echo* and *Telstar* demonstrated the feasibility of communications satellites that orbit the Earth at the same rate that the planet spins, staying above the same place on Earth.

By 1963, when Robert Wilson had begun work at Bell Labs, the Communications Satellite Corporation, abbreviated Comsat, had been created, and Bell Labs' parent company, the American Telephone and Telegraph Corporation, was legislated out of the international communications-satellite business to avoid a monopoly. So Bell Labs had this superb horn antenna and maser amplifier that were no longer needed for communications testing, and Penzias and Wilson decided to use them for radio astronomy.

To their surprise, Penzias and Wilson detected, in 1964, microwave signals that came from all directions equally, wherever they pointed the antenna. Even a seemingly empty part of space emitted the persistent radiation, and it continued unabated for one entire year, through all the four seasons as the Earth orbited the Sun. And the unexpected signals had no dependence on the location of the Sun, or for that matter the Milky Way or any other place in the sky.

The brightness of this radiation corresponded to a temperature of about 3 kelvin, just three degrees above absolute zero, the coldest temperature possible. Three degrees wasn't very much, but it was more than expected. The atmosphere could be ruled out, since its brightness would be proportional to the thickness of the atmosphere and vary with direction – more toward the horizon and less toward the zenith. Extraterrestrial radio sources could be eliminated as well, for they would also produce a directional variation, and the unexpected signals came from places where there were no known cosmic radio sources.

A pair of pigeons was initially suspected. The messy birds had covered the inside of the horn with their white droppings, but the unexpected radiation remained after Penzias and Wilson had disposed of the pigeons and cleaned up their "white noise." They just couldn't understand what was causing the mysterious signals.

Penzias and Wilson didn't know what to do, until one day Penzias happened to be talking to Bernhard Burke, a radio astronomer then at the Carnegie Institution of Washington, and mentioned their unexplained signal. Burke recalled hearing about theoretical work involving pervasive, cosmic radio radiation by P. James E. Peebles at Princeton University, who had been asked by his Princeton colleague Robert H. Dicke to investigate the possible significance of such radiation. So Burke called Dicke who sent him a copy of Peeble's calculations. They indicated that the Universe should be filled with a relic radiation of the Big Bang, cooled to about 10 kelvin and stretched into microwaves by the expansion.

It was not surprising that Dicke was one of the first to realize the importance of this radiation, for he had placed limits on its temperature twenty years earlier and was preparing to measure it at about the same time that Penzias and Wilson unexpectedly made their discovery. From 1941 to 1946 Dicke was involved in war-related radar work at the Radiation Laboratory of the Massachusetts Institute of Technology, where he invented a sensitive microwave detection system designed to measure return echoes when a radar beam is bounced back from an enemy plane or ship. Known today as the Dicke radiometer, it is still in use to measure the intensity of microwave signals. In 1946, he used the instrument to measure the absorption of radar signals by atmospheric water vapor, and in the process set an upper limit of 20 kelvin to the temperature of the "radiation from cosmic matter."

At Dicke's suggestion, two younger scientists at Princeton, Peter G. Roll and David T. Wilkinson, began construction, in 1964, of a microwave antenna and receiver to measure the relic radiation. But before Roll and Wilkinson had a chance to get any results, they learned that they had been scooped by the Bell Labs duo.

So Dicke and his coworkers visited Penzias and Wilson at their nearby field station in Holmdel, New Jersey, located just 40 kilometers (25 miles) from Princeton, and the two groups agreed to the publication of two adjacent letters in the *Astrophysical Journal*. The Bell Labs scientists would report on their discovery observations, while the Princeton group would write about their possible theoretical consequences.

In their letter, published in the 1 July 1965 issue, Penzias and Wilson carefully avoided any discussion of the cosmological implications of their discovery, since they had not been involved in that kind of work and thought their measurements were independent of the theory and might outlive it. Their publication, with the modest title "A Measurement of Excess Antenna Temperature at 4080 MHz," was limited to a discussion of an external signal, also called noise, that came from all directions with an antenna temperature of roughly 3.5 kelvin at a wavelength of 7.35 centimeters, or a frequency of 4080 MHz.

The companion paper by Dicke, Peebles, Roll and Wilkinson, which had a more ambitious title of "Cosmic Blackbody Radiation," argued that the ubiquitous and unvarying noise was that of a thermal (blackbody) cosmic background, a relic of the hot Big Bang. And sometime between the submission and publication of both papers, Walter Sullivan obtained copies and announced the discovery on the front page of the *New York Times*. The headline, on 21 May 1965, read: "Signals Imply a 'Big Bang' Universe," which told it all, and also helped Penzias and Wilson realize the possible importance of what they had found.

It was the faint, cooled after-glow of the Big Bang, now known as the three-degree cosmic microwave background radiation. Not only did it conclusively confirm the existence of a Big Bang, but it also suggested that the early history of the expanding Universe could still be visible. As we look further into space and back into time, we might investigate the entire evolution of the observable Universe, even detecting the places where galaxies form. And behind it all is the three-degree cosmic microwave background radiation, which is why it is known as a background. So it's not surprising that Penzias and Wilson were awarded the 1978 Nobel Prize in Physics for this incredible discovery.

Predicting the Cosmic Microwave Background Radiation

As it turned out, the Princeton explanation of the relic background radiation was not altogether new. It was predicted by two American scientists Ralph A. Alpher and

Robert Herman in 1948, following some ideas of their advisor in graduate school, the Russian-born American physicist George Gamow. After correcting some numerical errors and reformulating Gamow's description of the expanding, evolving Universe, Alpher and Herman concluded that the Big Bang radiation should still be around. Though dramatically cooled by the expansion, its present temperature was still above absolute zero, at about 5 kelvin according to their calculations. That's awfully close to the three-degree value observed in 1965, but back in 1948 nobody attempted to observe the relic radiation. Alpher and Herman subsequently talked to radio astronomers about the possibility, but the signal was thought to be too weak to be detected within the confusing noise from other sources.

Nevertheless, Alpher, Gamow and Herman were pretty upset when their earlier work was not mentioned in the 1965 papers describing the discovery and explanation of the cosmic microwave background radiation. Although Dicke had attended Gamow's talk at Princeton on the primeval fireball, either Gamow didn't mention that the radiation would still be present or Dicke forgot about it. In 1964 Peebles had submitted a paper to the *Physical Review*, estimating a current background radiation temperature of about 10 kelvin, but it was rejected on the grounds that Gamow, Alpher and Herman had arrived at similar conclusions sixteen years earlier. The historical record was corrected in review articles after the discovery, when the Princeton physicists acknowledged their oversight. This controversy over appropriate credits could account for the fact that no Nobel Prize was given for the prediction of the three-degree cosmic microwave background radiation, though one was awarded for its discovery.

As pointed out by Dicke and his colleagues in 1965, the radiation would have a blackbody spectrum if it were the cooled relic of the primeval fireball, and measurements of this spectrum provided the convincing proof of its origin in the Big Bang.

Blackbody Spectrum

In its early "fireball" stage, the Universe was very small and compact, with an exceptionally hot temperature in excess of 10 billion kelvin. At these high temperatures, any atoms would be torn apart into their constituent fundamental particles, the protons, neutrons and electrons, and the radiation would frequently interact with them. So everything would have the same temperature, achieving a thermal equilibrium. Later, when the Universe thinned out and cooled by expanding into a greater volume, the matter and radiation quit interacting, going their separate ways. But the radiation would have retained its thermal nature, characterized by a single temperature.

A perfect thermal radiator is known as a blackbody, which absorbs all thermal radiation falling upon it and reflects none – hence the term "black." The blackbody reradiates the incident energy at a wide range of wavelengths, but with an intensity that is greatest at certain wavelengths that depend on the temperature.

Once the radiation intensity is measured at one wavelength, a determination of the intensity at other wavelengths will establish if it is a blackbody. This is because the blackbody spectrum, or the distribution of its radiation intensity with wavelength, is described by a straightforward formula, first derived by the German physicist Max Planck in 1901.

Planck's derivation included his proposal that radiation energy is not a continuous, wave-like thing, similar to flowing water, but comes in individual packets, which he called quanta. He was awarded the 1918 Nobel Prize in Physics for this discovery of energy quanta. They account for the unique asymmetric shape of the blackbody radiation spectrum.

Planck's formula indicated that the intensity of blackbody radiation peaks at a certain wavelength, dropping precipitously at shorter wavelengths and falling off gradually at longer ones (Fig. 4.16). The wavelength of maximum intensity is inversely proportional to the temperature. And regardless of the peak's location, a hotter blackbody is brighter at all wavelengths and radiates more total energy than a colder one.

The expansion of the Universe preserves the blackbody spectrum of the radiation for all time. No process can destroy its shape, but the location of maximum intensity will stretch to longer and longer wavelengths as time goes on and the radiation gets colder. In the present epoch, with a temperature of about 3 degrees above absolute zero, the blackbody radiation intensity peaks at a wavelength of 0.001 meters, or 1 millimeter. Unfortunately, the Earth's atmosphere absorbs cosmic radiation at this short wave-

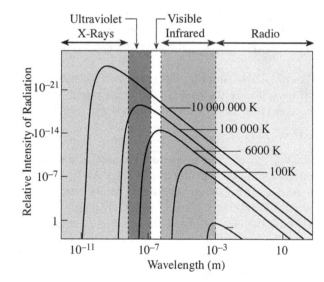

FIG. 4.16 Thermal radiation at different temperatures
The spectral plot of thermal radiation intensity as a function of wavelength depends on the temperature of the gas emitting the radiation. At higher temperatures, the most intense emission occurs at shorter wavelengths, and the thermal radiation intensity becomes greater at all wavelengths. A hot gas with a temperature of a million kelvin emits most of its radiation in X-rays, while the emission peaks at visible wavelengths near 5,780 kelvin, the temperature of the Sun's disk. At much lower temperatures, of just a few kelvin, the thermal radiation is most intense at radio wavelengths.

length, and the initial checks of the expected blackbody spectrum had to be performed at longer wavelengths that reach the ground.

Penzias and Wilson's 1965 measurements were made at a single wavelength of 7.35 centimeters, obtaining a temperature of 3.0 kelvin with an uncertainty of about one kelvin. The first confirmation of the radiation's blackbody spectrum came several months later in early 1966, when Peter G. Roll and David T. Wilkinson reported the measurement of the same temperature at a wavelength of 3.2 centimeters (Fig. 4.17). They were already constructing an instrument to search for the radiation in 1965, and

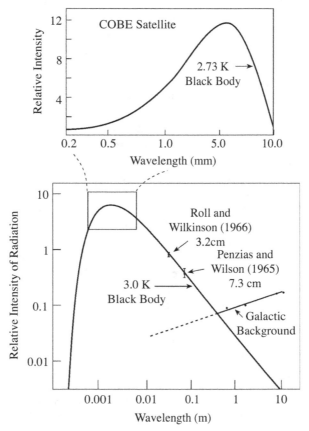

FIG. 4.17 Spectrum of the cosmic microwave background radiation Pioneering measurements of the brightness of the cosmic microwave background radiation, by Arno A. Penzias and Robert W. Wilson in 1965 and Peter G. Roll and David T. Wilkinson in 1966, at 7.35 and 3.2 centimeters wavelength respectively, compared to the expected spectrum of a three-degree blackbody and radiation from our Galaxy (*bottom*). The full spectrum at millimeter wavelengths (*top*) was obtained from instruments aboard the *COsmic Background Explorer*, abbreviated *COBE*, satellite in late 1989. The solid line corresponds to a thermal radiator, or blackbody, with a temperature of 2.726 ± 0.010 kelvin.

quickly realized that a determination of its blackbody spectrum would be crucial proof of its origin in the Big Bang.

Of course, measurements at only two wavelengths are insufficient to determine the shape of the spectrum. But within a year, a blackbody spectrum was indeed verified by numerous measurements at a wide range of wavelengths.

The confirmation of a blackbody spectrum convinced most astronomers that the three-degree cosmic microwave background radiation is left over from the dense, early stages of the expanding Universe, and finally persuaded Wilson that the relic radiation was indeed of cosmological importance. As he later recalled, "we felt that, at least until they [the Steady-State theorists] had a chance to think about our results, we shouldn't go out on a theoretical limb that we couldn't support. For me, the last nail in the coffin of the Steady-State theory wasn't driven in for quite a while – not until the blackbody curve was really verified."[49]

The definitive spectral measurements were made from above the Earth's atmosphere using NASA's *COsmic Background Explorer*, abbreviated *COBE*. Originally designed to be launched from the *Space Shuttle*, *COBE* was at least partly delayed when the shuttle *Challenger* exploded soon after launch in 1986. After reconfiguration, *COBE* was finally launched on 18 November 1989 aboard a Delta rocket.

On 13 January 1990, less than two months after *COBE* went into orbit but a quarter of a century after the discovery of the cosmic radiation, John C. Mather reported the combined results of millions of *COBE* spectral measurements at an American Astronomical Society meeting near Washington, D.C.. The spectrum fit the Planck blackbody curve with a precision of one part in 10,000 (also shown in Fig. 4.17), establishing a temperature of precisely 2.726 kelvin, with an uncertainty of 0.010 kelvin. The presentation caused the audience to break into a standing applause, which you usually see only at the end of a beautiful performance of music.

Such a spectrum could not have happened in the Universe as it is now. Matter currently has a very different temperature than the background radiation. And to put it another way, the observed spectrum is proof that the observable Universe did expand from a very hot, dense state in the past, when matter and radiation were in thermal equilibrium and at the same temperature.

As Smooth as Silk

The thing that alerted astronomers to the importance of the background radiation was its equal brightness wherever one looked, indicating that it uniformly fills all of space. This spatial isotropy satisfied one of the basic tenants of modern cosmology, the cosmological principle, which asserts that except for local irregularities the Universe presents the same aspect from every point.

But the radiation seemed too uniform. When a smoothly varying change caused by the Earth's motion was removed, the *COBE* instruments could detect no regions brighter than others down to 0.0003 kelvin, or one part in 10,000, on angular scales from minutes of arc to 180 degrees.

Yet, the background radiation, itself a relic of time, ought to have waves and ripples in it, which provided the seeds from which galaxies grew. Somewhere in all that smoothness, there ought to be departures from silky uniformity, like the folds in a courtesan's robe. They must have acted as a template or blueprint, encoding the information required to explain the subsequent evolution of the Universe, the astronomical equivalent of the human genome. Otherwise we wouldn't have the lumpy distribution of galaxies that is now observed.

Cosmic Ripples

To the relief of many astronomers, George Smoot and his colleagues announced measurements of the temperature fluctuations in 1992, using four years of data gathered by *COBE*. After subtracting out the known microwave emission of the Milky Way and using mathematical averaging techniques on about 100 million observations, the *COBE* team found that the temperature varies ever so slightly over large angular sizes on the plane of the sky (Fig. 4.18). The sensitive instrument just barely detected minute temperature differences no larger than a hundred-thousandth, or 10^{-5}, of a degree.

When the colored, oval-shaped *COBE* map was unveiled, the fanfare was incredible. While discussing it at the 23 April 1992 meeting of the American Physical Society, Smoot compared it to a mystical experience. "If you're religious," he said, "this is like looking at God." This pretentious remark appeared on the front page of every major newspaper the next morning. Moreover, Smoot's home institution, the Lawrence Berkeley National Laboratory, had issued a press release two days earlier, including Stephen Hawking's remark that the *COBE* result was "the scientific discovery of the century, if not of all time." That was stretching the truth more than a little, but the overall effect made Smoot an instant celebrity, while perhaps overlooking the contributions of other members of his team.

COBE was pushed to the limits of its sensitivity, with evidence that wasn't quite definitive, mainly because it mapped the cosmic radiation with coarse angular resolution greater than 7 degrees. In the subsequent decade, more than 20 experiments were therefore carried out from the ground and balloon platforms, bringing the temperature fluctuations into shaper focus with angular resolutions as fine as a few minutes of arc, comparable to the size of galaxy concentrations. They included the ACBAR instrument near the South Pole, the BOOMERanG instrument on balloons above Antarctica, and the CBI radio telescope in the Chilean Andes. The acronyms are enough to boggle your mind: ACBAR for the Arcminute Cosmology Bolometer Array Receiver, BOOMERanG for Balloon OBservations of Millimetric Extragalactic Radiation and Geophysics, and CBI for the Cosmic Background Imager.

The BOOMERanG, for example, mapped about 2.5 percent of the Cosmic Microwave Background using a balloon-borne telescope that circumnavigated Antarctica. Already in 2000, and again in 2001, it obtained detailed measurements of the characteristic angular sizes of the temperature variations, which were attributed to primordial sound waves. Most of the fluctuating power was emitted at an angular size of about 1 degree, and this indicated no detectable curvature of space, which would have acted as a lens and made the expected size appear larger or smaller after traveling across 14 billion light-years. In other words, the Universe could be described by Euclidean geometry, without the space curvature employed in *General Relativity*.

But the ground-based and balloon experiments only glimpsed small portions of the sky for a limited time, so there was a possibility that they might not be truly

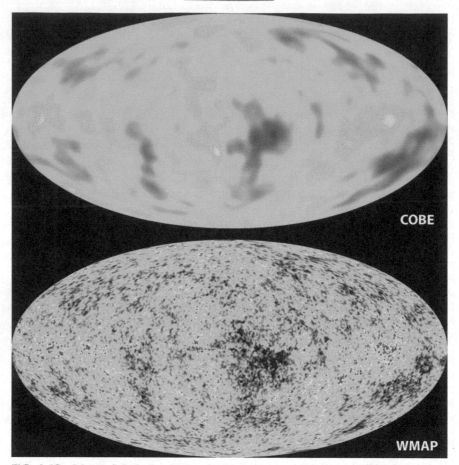

FIG. 4.18   Maps of the infant Universe Views of the three-degree cosmic microwave background radiation emitted from the Universe in its infancy, just 380,000 years after the Big Bang. These images cover the full sky with contributions of the Milky Way removed. They were taken by instruments aboard the *Cosmic Background Explorer,* abbreviated *COBE,* satellite in 1996 (*top*) and by the *Wilkinson Microwave Anisotropy Probe,* or *WMAP* for short, satellite in 2003 (*bottom*). The *WMAP* image brings the *COBE* picture into sharp focus. The microwave background radiation took more than 13 billion years to reach the two satellites. The temperature fluctuations range up to 0.0002 kelvin above and below the average value. Darker regions are cooler and lighter ones are hotter. These fluctuations subsequently grew to become galaxies. (Courtesy of the NASA/*COBE* and NASA/*WMAP* Science Teams.)

representative of the whole sky. It was therefore time for another satellite experiment that would scan the entire sky without the confusion of microwave radiation from the atmosphere and ground. This time, the spacecraft wouldn't just detect the cosmic ripples. It would instead determine their distribution and characteristic sizes, filling in the gaps between the small features seen with other instruments and the large ones detected by *COBE.*

At about this time, NASA was faced with intense public scrutiny, having launched the *Hubble Space Telescope* with a faulty mirror, lost the billion dollar *Mars Observer* spacecraft, and blown up at least one *Space Shuttle*. So NASA administrator Daniel S. Goldin decided to cut the losses, and to begin doing things "better, faster and cheaper" using mid-sized rockets to launch inexpensive missions within a year or two of approval.

David T. Wilkinson, at Princeton, joined Charles L. "Chuck" Bennett at the Goddard Space Flight Center to create a small team of experts and design a spacecraft that could do the job within the budget cap of $70 million in 1994 dollars. The resultant *Microwave Anisotropy Probe*, or *MAP* for short, was approved in mid-1996 and launched on 30 June 2001, with a name change in early 2003 to *WMAP*, or *Wilkinson Microwave Anisotropy Probe* to honor Wilkinson after his death.

The most detailed and precise map yet provided for the background radiation was obtained from *WMAP* in 2003 (also shown in Fig. 4.18). Its instruments scanned the entire microwave sky for more than two years, with 100 times the sensitivity and 30 times the resolving power of *COBE*, recording features as small as 13 minutes of arc across. This time, there could be no doubt that the temperature fluctuations were real, and they could reliably tell us what the young Universe was made of, how dense it was, and what forces were at play at the time.

When combined with previous measurements, the *WMAP* instruments showed that the temperature variations are concentrated within certain angular sizes, which are displayed in an angular power spectrum, a plot of the relative strength of the hot and cold spots against their sizes (Fig. 4.19). The inflation model, of an accelerated growth in the earliest moments of the expanding Universe, predicts that this spectrum

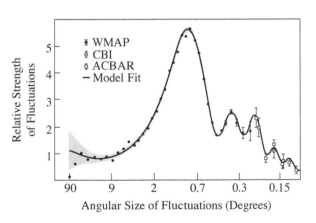

FIG. 4.19 Ripple data The angular fluctuation strength, or power, of the cosmic microwave background radiation is displayed as a function of its angular size. The solid line is the model that best fits the data (*solid dots*) from the *Wilkinson Microwave Anisotropy Probe*, abbreviated *WMAP*, obtained in 2003; the gray band represents uncertainties in the model. The first two data points (*left*) at the largest angular sizes are anomalously low in power, falling below the curve, and they are in disagreement with inflation models for the early Universe. Anisotropy data obtained by previous experiments are denoted by dots with error bars (*right*). The observed power spectrum has been compared to other astronomical observations and different theoretical models, providing estimates for the amount of dark matter and dark energy in the Universe (Fig. 5.14). (Courtesy of the NASA/*WMAP* Science Team.)

ought to be flat, without distinguishing features, but it is not. The observed ripples are supposed to be introduced after inflation terminated, and gravity caused regions of enhanced density to collapse in on themselves. But inflation models cannot explain the missing power at the largest angular sizes, and dark, unseen particles are required to help the process of collapse that accounts for the ripples.

The ratio of the heights of the first and second peak of the angular power spectrum has been used to determine the amount of "ordinary" matter that we are used to, the baryonic kind that makes up atoms. And when the height of the third peak was compared to those of the other two, scientists estimated the amount of dark, non-baryonic matter. They indicate that dark matter is five times more abundant than ordinary baryonic matter, which is consistent with other astronomical measurements, and that the combined gravitational pull of both kinds of matter is not enough to stop the expansion of the Universe in the future. But there are other unexpected things that pervade the apparent emptiness of space.

Chapter Five
The Fullness of Space

5.1 SPACE IS NOT EMPTY

Throughout history, the notion of emptiness has been inseparable from our explanations of the material world that we can see, touch and feel. To the ancient Greek philosopher Democritus, all matter consisted of elementary particles, which he called *atoms* from the Greek word for "indivisible." The space between the atoms was supposed to be empty, a nothingness without anything, so the atoms would have somewhere to go, to move into.

In modern times, physicists have shown that atoms are composed of smaller things. When bombarding gold foil with energetic particles in 1910, for example, the British physicist Ernest Rutherford was astounded to find that some of them bounced back while most did not. The returned particles had hit a tiny nucleus that is about 100,000 times smaller than the atom itself, and this means that an atom is mainly empty space, just as the room you are sitting in is largely empty.

And when looking out, instead of in, there is the seductive darkness of cosmic space. You can sense its fascinating allure, when night falls, "with the softness of a woman,"[50] or when the music of the night "unfurls its splendor," to "caress you" and "secretly possess you."[51] And in the words of Henry Wadsworth Longfellow:

The night shall be filled with music
And the cares that infest the day,
Shall fold their tents, like the Arabs,
And as silently steal away.[52]

When the day is done, we gradually become aware of another, quieter world. At twilight, fish rise to the surface of a lake and swallows emerge to dart through the air. Raccoons begin to forage and prowl about, bats leave their caves, and the moonlight primrose opens its petals, keeping them in blossom all night long.

There is music, poetry and mystery in the enfolding black of night, in the haunted chambers of its darkness. The Swiss artist Paul Klee wrote about the emptiness of the night,[53] but he also believed in its power, portraying light that comes from darkness itself. In his *Nocturnal Flowers* (Fig. 5.1), plants take root from all sides of the composition, and converge toward a black central Sun.

FIG. 5.1 Nocturnal flowers This watercolor, painted in 1918 by Swiss artist Paul Klee (1879–1940), contains flowers, trees, and exotic plants that grow from all four sides of the painting toward a black central Sun, while the entire landscape remains submerged in a sea-like blue. You could view the painting from any side. In the perspective shown here, a flower is growing down from the top, a mysterious red tower rises from the bottom edge, and from the left, star-like flowers are growing out toward a silver crescent Moon. Klee saw things differently from many of us. During a trip to a Swiss nature reserve, a friend lost sight of Klee, eventually finding him sitting on a rock. Klee had been staring at the ground, and when he raised his head, he announced, "I've just had a long conversation with a snake."[c] Watercolor 0.17 × 0.17 m. (Courtesy of the Folkwang Museum, Essen.)

Cerius, a fragrant, night-blooming cactus, produces a large white flower with a yellow, Sun-like center, but it blooms on only one night a year, fading the next morning. The Spanish call it *reina de la nouche,* or "queen of the night", but its reign is short.

Empty space also provides a transparent structure to the world, giving form and shape to material things, defining their place and freeing them from their surroundings. A sculptor, for example, creates an object as much by removing material as by adding to it. And the French painter Henri Matisse reduced many of his works to the

[c] Paul Klee (1879–1940), In Jean-Louis Ferrier, *Paul Klee,* Pierre Terrail Editions, Paris, 1998, pages 64, 65.

FIG. 5.2 **The joy of life** French artist Henri Matisse (1869–1954) created this revolutionary painting about a century ago, in 1905 – 6. With its extreme simplicity, the painting helped initiate a new perspective in art, one in which the artist concentrates on what to omit rather than what to include. Figures have been pared down to flowing lines and color, surrounded by large empty places. It is the missing parts and the vacant regions that provide a different way of looking at the world. Oil on canvas. 1.74 × 2.38 m. (Courtesy of the Barnes Foundation, Merion, Pennsylvania.)

bare minimum (Fig. 5.2). Rather than deciding what to include, he thought about what ought to be omitted, removing everything that could be left out and replacing them with emptiness and spatial depth. And another French painter, Rene Magritte, proposed that anything we focus our vision on hides something else we might want to see. His paintings show us the irony of our selective vision by making us look at images we never see in nature, such as people falling like raindrops from the sky.

Moreover, the unseen chambers of nearby space are not empty. The slow drift of floating clouds, or the sight of airplanes supported by their motion, proves that there is something substantial surrounding us. Hawks circling above a warm meadow are getting free rides in the rising currents of hot air, sometimes rising so abruptly that it looks as if they were lifted and jerked up by strings. And we can sense the touch of the wind on a stormy day, or feel the air against our skin on a cold night.

Travel beyond our atmosphere to the right place just outside the spinning Earth, and you will be bathed in perpetual sunlight without any night at all. And radiation isn't the only thing out there. The Sun's tenuous outer atmosphere is streaming away into seemingly empty space, enveloping the Earth and other planets in a continuous, invisible wind of electrons and protons. So the space between the planets, once thought to be a tranquil, empty void, is swarming with hot, charged invisible pieces of the Sun.

Further out, the space between the stars, which looks like an empty black void, is filled with cold, invisible atoms of hydrogen that have been around since the beginning of time. And stars other than the Sun are blowing themselves away into interstellar space, either slowly or explosively. We see some of this interstellar material when the brightest stars light up the space around them, even in neighboring galaxies.

The vast reservoir of supposedly empty space is also filled with dark matter, which seems to dominate the dynamics of visible stars, and mysterious dark energy that is apparently pushing the expanding Universe out at an ever-increasing speed.

The vacuous nothing of space might even be the ultimate source of everything that exists today, both observed and unobserved, explaining how the Cosmos went from nothing at all to a marvelous something. All the mass and energy in the observable Universe could conceivably have been made from the seemingly empty nothingness of space, perhaps during a rapid, brief ballooning out immediately after the Big Bang. Or it might be miniscule vibrating strings that made everything out of nothing.

But to let our story unfold in the usual way, from the Earth to the wider space beyond, let us begin with the space between the planets, which is filled with a ceaseless wind blowing from the Sun.

5.2 WE LIVE INSIDE THE EXPANDING SUN

Just half a century ago, most people visualized our planet as a solitary sphere traveling in a cold, dark vacuum around the Sun. But we now know that the Sun is forever blowing itself away in a solar wind that moves past the planets and engulfs them.

People think that the Earth revolves around the Sun. But it's better to think that the Earth moves within the Sun. For we are actually living in the outer part of the Sun, and in this sense our planet is in the Sun rather than separated from it.

Unlike any wind on Earth, the solar wind is mainly composed of electrons and protons set free from the Sun's abundant hydrogen atoms, but it also contains heavier ions and magnetic fields. They all expand and flow away from the Sun, forming a perpetual solar wind that was suggested by terrestrial auroras, inferred from comet tails, implied by theoretical considerations, and fully confirmed by direct measurements from spacecraft.

The first suggestive evidence for the fullness of interplanetary space came not from looking at the Sun or other stars, but instead at our home planet where curtains of green and red light dance and shimmer across the polar skies, far above the highest clouds. This light is called the aurora after the goddess of the rosy-fingered dawn, a designation that has been traced back to Galileo Galilei.

At the turn of the century, the Norwegian physicist Kristian Birkeland argued that the Earth's magnetism focuses incoming electrons to the polar regions where they produce auroras. This is because magnetic fields create an invisible barrier to charged particles, causing them to move along their magnetic conduits rather than across them. Although Birkeland's proposal was not widely appreciated in his time, we now know that he was right. The Earth's dipolar magnetic field guides energetic electrons into the upper atmosphere near the magnetic poles where they collide with oxygen and nitrogen molecules, causing them to glow like a cosmic neon sign.

Comets provided another clue to the existence of the Sun's expanding outer atmosphere; somewhat like a wet finger placed in a terrestrial wind to show it is there. When a comet is tossed into the inner Solar System, the Sun's warmth vaporizes dirty ice on the comet's surface, forming two tails that always point away from the Sun. One is a yellow tail of dust, pushed away from the Sun by the pressure of sunlight. The other is a straight ion tail, shining in the electric-blue light of ionized particles. As proposed by the German astronomer Ludwig Biermann in 1951, the straight ion tail is pushed away from the Sun by a ceaseless, electrified flow of charged solar particles.

At about the same time, it was realized that the outer atmosphere of the Sun, just above its visible disk, has an unexpectedly high temperature of a few million kelvin. Visible during a total solar eclipse (Fig. 5.3), and known as the solar corona, the gas is so hot that it overcomes the Sun's gravitational pull and accelerates away to supersonic speed, carrying its searing heat and electrified particles to the Earth and beyond.

But the corona is so rarefied that we look right through it, just as we see through the Earth's clean air. That is the reason that the space outside the Sun looks so empty; it's just exceedingly tenuous, far less substantial than a terrestrial breeze or even a whisper. The density of the solar wind is so low that if we could go out into space and put our hands in it, we would not be able to feel it. And even at its origin near the visible Sun, the pressure of the solar wind is no more than the pressure under the foot of a spider.

FIG. 5.3 Gossamer corona The Sun's expanding outer atmosphere, the corona, as photographed during the total solar eclipse of 26 February 1998 from Oranjestad, Aruba. To extract this much coronal detail, several individual images, made with different exposure times, were combined and processed electronically in a computer. The resultant composite image shows the solar corona approximately as it appears to the human eye during totality. Note the fine rays and helmet-shaped streamers that extend far from the Sun and correspond to a wide range of brightness. (Courtesy of Fred Espenak.)

Modern spacecraft have also been used to observe the X-rays emitted by the million-degree corona. The hottest and densest X-ray emitting material is held near the Sun within magnetic loops (Fig. 5.4). But some of the million-degree gas escapes, moving at high speeds from the dark coronal holes that are nearly always present at the Sun's poles.

The charged particles retain most of their million-degree heat as they travel to the Earth, but since the hot gas is so rarefied, spacecraft and astronauts are not burned up when entering the solar wind. There aren't enough particles around to fry anything, not even an egg.

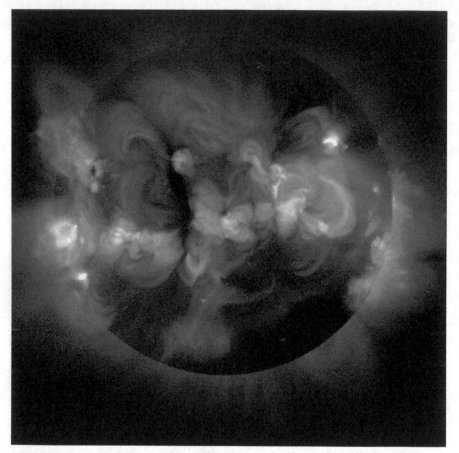

FIG. 5.4　The Sun in X-rays Ionized gases at a temperature of a few million kelvin produce the bright glow seen in this X-ray image of the Sun. It shows magnetic coronal loops that thread the outer solar atmosphere, known as the corona, and hold the hot gases in place. The brightest features are called active regions and correspond to the sites of the most intense magnetic field strength. The Soft X-ray Telescope aboard the Japanese *Yohkoh* satellite recorded this image of the Sun's corona on 1 February 1992, near a maximum in the 11-year cycle of solar magnetic activity. (Courtesy of Gregory L. Slater, Gary A, Linford, and Lawrence Shing, NASA, ISAS, the Lockheed-Martin Solar and Astrophysics Laboratory, the National Astronomical Observatory of Japan, and the University of Tokyo.)

Nevertheless, small things can have a big effect. Although it is exceedingly thin, the solar wind is powerful enough to mold the outer edges of the Earth's magnetic field into an asymmetric shape, like a teardrop falling toward the Sun. We reside within this magnetic cocoon, known as the magnetosphere, which diverts most of the solar wind around our planet at a distance far above the atmosphere, like a rock in a stream or a windshield that deflects air around a car. The magnetosphere thus protects humans on the ground from possibly lethal particles carried in the Sun's wind.

When out alone in deep space, working on a space vehicle or walking on the Moon or Mars, astronauts can be exposed to powerful gusts in the solar wind, with no place to hide. Exceptionally energetic wind particles, hurled out by explosions on the Sun, can even pass through satellite walls and an astronaut's eyelids, striking the retina and making an astronaut's eyes glow inside. During future space missions, solar astronomers will keep careful watch over the Sun, providing timely warning of solar explosions and alerting astronauts to come in from outside activities to seek shelter in metal sanctuaries or underground caves on the Moon or Mars.

5.3 THE SUN CURVES NEARBY SPACE

Space and Time are Joined Together

Nothing likes to be crowded too close together, from electrons to people. To paraphrase something the great American musician, Wynton Marsalis, once said, sometimes the closer together you get, the farther apart you become, like man and woman. A pressure is created that resists the confinement, opening up the space between objects and keeping them at a distance.

You might have noticed an unrestrained sense of freedom when swimming far from a shore where there is plenty of "empty" space and no restraining boundaries. There is nothing out there to give us our bearings. As the American folk singer, Bob Dylan, wrote: "She never stumbles, she's got no place to fall."[54]

Empty space can make us temporarily forget the relentless onward march of time. Past and future become lost when sailing far from shore in the endless emptiness of the ocean, with its blue-green surface melding into a silvered sky without a seam or horizon.

Yet, time is always there, establishing some unavoidable constraints. It has a beginning and an end, always flowing forward in just one direction, from the past to the future. As movies and old family photographs have taught us, what you see is relative to our position in both space and time.

And ongoing time means change. Everything we know, from our closest friends to the most distant galaxies, is always growing older as the clocks tick away, marking the inevitable passing from one moment of time to another.

Quite ordinary motions apparently disturb the sheet of time, providing a pattern to our daily lives. The spinning Earth marks out the passing days from sunrise to sunset and back to sunrise again. And time really seems to fly when you're moving fast, rushing to the next appointment or dancing all night.

We often join our concepts of space and time without noticing that we are doing so. When you make an appointment to meet someone, you specify where and when, or the place and time of the meeting. And although the distance to a remote location is often specified by how far away it is, the distance is sometimes described by how long it takes to get there.

Space and time were linked and joined together in the *Special Theory of Relativity*, anticipated by the French mathematician Henri Poincaré and fully developed by Albert Einstein in 1905. It combines four dimensions, three for space, and one for time, with the velocity of light thrown in to give time the units of space, forming an inseparable entity called space-time.

In this description, two events don't have a uniquely defined separation in either space or time. Instead, they are separated in space-time. So, the concepts of space and time are interwoven. As Einstein's former teacher, Hermann Mikowski, declared in 1908: "Henceforth space by itself, and time by itself, are doomed to fade away into mere shadows, and only a kind of union of the two will preserve an independent reality."[55]

The *Special Theory of Relativity* supposes light is always moving at a constant speed with the precise velocity of 299,792.458 kilometers per second, traveling forever in empty space, never stopping or slowing down and never coming to rest. Moreover, nothing outruns light; it's the fastest thing around.

The unvarying speed of light was first demonstrated in a famous experiment, performed by the American physicist Albert A. Michelson, assisted by his friend the chemist Edward W. Morley, in 1887 (Focus 5.1).

According to *Special Relativity*, high-speed motion changes our everyday notions of space and time. We normally regard time as absolute and immutable, with nothing disturbing its relentless, steady tick. But for Einstein, time is relative and variable. In

---

FOCUS 5.1
The Michelson-Morley Experiment

Many experiments have been carried out to confirm the unvarying speed of light, but the most famous one was conducted in a basement laboratory at the Case School of Applied Science in Cleveland, Ohio in 1887, when the American scientists Albert A. Michelson and Edward W. Morley attempted to use an interferometer to precisely measure how the speed of light depends on the Earth's motion through space.

Scientists of that time firmly believed in an imaginary luminiferous ether, an invisible, frictionless, and unmoving medium that was supposed to permeate all of space. Its presence explained how light waves could travel at high speed through the apparent emptiness of space, providing the medium in which they propagate and vibrate.

As the Earth moves through the stationary ether, an ether wind should blow past the Earth in the direction of its motion. And the speed of light would vary, like a boat sailing with or against the wind, or a swimmer moving downstream or struggling upstream. But Michelson and Morley found that there was no detectable difference in the interference pattern produced when a beam of light was sent into the ether wind in the direction of the Earth's motion or directed at right angles to it. Moreover, there was no difference in the measured speed of light when the Earth was traveling toward the Sun and away from it half a year later.

So the experiment meant that there was no light-carrying ether. It also meant that the velocity of light is constant, exactly the same in all directions and at all seasons, and independent of the motion of the observer.

rapid travel, the rate at which time flows decreases, so clocks run slower. And lengths are diminished at high speed, shrinking in the direction of motion.

All is relative, even mass, which increases with the speed. That is why nothing can move faster than the speed of light. An object's mass becomes infinite at that speed, and nothing could push it faster.

But we never notice these effects. In the familiar low-speed world we live in, where motions are exceedingly slow compared to the speed of light, size and time do not depend on motion to any significant degree. The differences produced by moving at ordinary speed happen, but they are miniscule and go unnoticed, even in the fastest airplanes. You would age slower and prolong your life by only 0.001 seconds if you spent your entire lifetime orbiting the Earth in a jet airplane.

Substantial, noticeable differences in the passage of time only happen at speeds that are a sizeable fraction of the velocity of light. Such speeds are completely unattainable in everyday life, and far beyond anything you will ever experience. But they are quite common during violent activity found on the Sun and elsewhere in the Cosmos.

In his book *Einstein's Dreams,* the American physicist and author Alan Lightman has imagined what would happen at a place where everything moves at the velocity of light and time stands still, where "Raindrops hang motionless in the air. Pendulums of clocks float mid-swing. Dogs raise their muzzles in silent howls. Pedestrians are frozen on the dusty streets, their legs cocked as if held by strings," and "One sees lovers kissing in the shadows of buildings, in a frozen embrace that will never let go."[56]

As the name implies, the *Special Theory of Relativity* deals with "special" circumstances, involving reference frames in uniform, unimpeded motion and ignoring gravity. When significant sources of gravity are added to the theory, space is no longer empty. Space is given a curvature that provides it with physical reality.

Mass-Energy Bends Space-Time

Gravity is always pulling us down, but its attraction decreases with distance. So when someone is out in space, far from the Earth's gravity, the pull is diminished. Astronauts consequently float practically weightless in their space vehicles, and their spinal cords lengthen about 7 centimeters without the downward pull of gravity at the Earth's surface. But their muscles also weaken from the lack of activity, and the change can be permanent. As the saying goes, if you don't use it, you lose it.

The great English physicist Isaac Newton described the action of gravity as a force that explains the motions of the planets and is transmitted instantaneously over vast distances of space. Then in 1915 – 16, Albert Einstein abolished the force of gravity and rewrote the theory in his *General Theory of Relativity,* explaining gravity in terms of geometry. He replaced the force of gravity with the warping of space and time around a massive body (Figs. 5.5, 5.6).

In effect, space has both content and shape. The curved shape is molded into space by its content, the massive objects. That bending, twisting and distortion of space is gravity. But such effects are only noticeable in extreme conditions near a very massive object, and the differences between the Newton and Einstein theories are indistinguishable in everyday life.

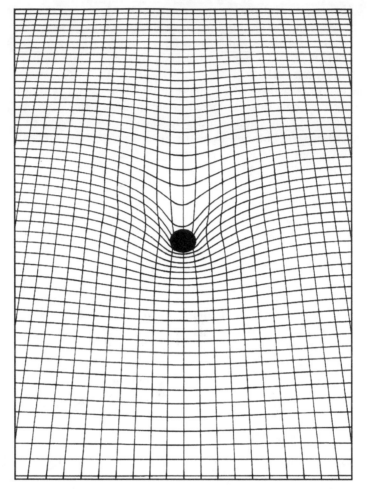

FIG. 5.5 **Space curvature** A central, massive, black disk creates a curved indentation upon the Euclidean space that describes a world that is without matter. Notice that the amount of space curvature is greatest in the regions near the object, while further away the effect is lessened.

In 1915, the year that Einstein first thought about the curvature of space, the Babelsberg astronomer Erwin Freudlich called attention to long unexplained anomalies in the motion of Mercury. Instead of returning to its starting point to form a closed ellipse in one orbital period, Mercury moves slightly ahead in a winding path that can be described as a rotating ellipse. As a result, the point of Mercury's closest approach to the Sun, the perihelion, advances by a small amount, 43 seconds of arc per century, beyond that which can be accounted for by planetary perturbations using Newton's theory of gravitation.

FIG. 5.6 Dominant curve The Russian artist Wassily Kandinsky (1866–1944) regarded this 1936 painting as one of his most important works. As the title suggests, a broadly designed curve dominates its surroundings, like a path to the heavens. Block-like steps in the lower right reinforce the ascent, as if one were climbing to the three black circles at the upper right and to the rectangular tablet at the upper left corner, which seems to decipher the Cosmos with signs. The colored disks and numerous smaller curves also seem to portray outer space, as in many of Kandinsky's other works. Oil on canvas. 1.3 × 1.95 m. (Courtesy of the Solomon R. Guggenheim Museum, New York.)

Although discovered in 1854, this anomalous twist in Mercury's motion remained unexplained until Einstein read Freudlich's article and solved the problem, by showing that the curvature of space in the neighborhood of the massive Sun guides the motion of Mercury. The planet is traveling as straight as it can along an invisible curved track in space-time, like a bird flying along currents of air or a ball spinning in a roulette wheel, making Mercury overshoot its expected perihelion position by precisely the observed amount.

Remembering his discovery of the explanation, Einstein wrote "for a few days, I was beside myself with joyous excitement," and confessed that it "had given him palpitations of the heart, with a feeling that something actually snapped in him."[57]

So the great theoretician was deeply motivated by the real world, testing his ideas against observable results. And like any good scientist, he realized that his new theory needed to be verified by definitive predictions of other consequences. So in the very paper that explained Mercury's anomalous motion, Einstein showed that the theory has as a consequence a bending of light rays due to gravity.[58]

Einstein noticed that light is heavy, or that it has energy and therefore mass, and that the Sun's gravity will pull the light toward it. In fact, the Sun's deflection of light had been inferred more than a century earlier. In the first edition of his *Opticks,* published in 1704, Newton already speculated that massive bodies might bend nearby light rays, and the English natural philosopher Henry Cavendish calculated the amount of the Sun's light deflection in 1780, under the assumption that light is composed of material corpuscles that are attracted by the Newtonian force of gravity.

Einstein confirmed this result more than a century later, showing in 1911 that a light ray passing by the Sun's visible edge would be deflected by 0.87 seconds of arc. Freudlich arranged to check this prediction of the bending of starlight during an eclipse visible in central Russia on 21 August 1914, but World War I (1914 – 1918) intervened. In his 1915 paper explaining Mercury's perihelion motion, Einstein corrected his first calculation, showing that the Sun's curvature of space increases the expected deflection by a factor of two, to 1.75 seconds of arc for a light ray grazing the Sun's edge. In this famous prediction, starlight follows a path in space-time that has been curved by the Sun; it acts like a lens and deflects the light of stars that are about to pass behind it. This pushes the apparent positions of nearby stars further apart and makes their separations look bigger.

Although the theory had been proposed by a German professor during a devastating World War, interest ran high in Britain, and funding was obtained for two joint Greenwich Observatory – Royal Society expeditions to measure the deflection of starlight around the Sun at the first opportunity, during the total solar eclipse on 29 May 1919, from a cocoa plantation on Principe Island off the coast of West Africa and from Sobral, Brazil.

In the last evening before sailing to Principe, E. T. Cottingham, who was to accompany the English astronomer Arthur Eddington, asked Sir Frank Dyson, the Astronomer Royal, what would happen if they found *twice* Einstein's predicted deflection. Sir Frank replied, "then Eddington will go mad and you will have to come home alone."[59] When subsequently hearing of the success of the eclipse expeditions, Einstein was similarly asked what he would have thought if the experiments had not confirmed his prediction, Einstein replied, "then I would have been sorry for the dear Lord, for the theory is correct."[60]

Eclipse conditions were favorable at both locations. Measurements were made of the difference in stellar positions on photographs taken during the eclipse and those obtained long before when the same region of the sky was nowhere near the Sun. The results, presented by Eddington to the Royal Society of London on 6 November 1919, indicated light deflections equivalent to 1.61 and 1.98 seconds of arc at the edge of the Sun using the Principe and Sobral observations respectively. Both values were roughly twice that predicted from Newtonian theory, and consistent, within the measurement uncertainties, with Einstein's prediction of 1.75.

The successful test made Einstein famous, practically overnight. The 7 November 1919 headline of the London *Times* read "Revolution in Science – New Theory of the Universe – Newtonian Ideas Overthrown," and news of the discovery hit the front pages of every important newspaper in the world.

The solar curvature of space has been measured with increasingly greater precision for nearly a century, confirming Einstein's prediction to one part in a thousand, or to the third decimal place. In one test, radio astronomers have combined observations of distant quasars from telescopes located on opposite sides of the world, turning the Earth into a gigantic interferometer that accurately measures the positions of radio sources as our line of sight to them nears the Sun. The change in position determines the amount of space curvature, as with the solar eclipse results for stars, but with much greater accuracy and without a total solar eclipse, since the Sun is a relatively weak interferometric radio source.

Another test of Einstein's theory of gravitation measures the time required for a radar signal to make a round trip between the Earth and a planet, or the time for a radio signal to travel from a spacecraft home. When the line of sight passes near the Sun, the radio waves travel along a curved path and take slightly longer to return to Earth. The measurements require extremely precise clocks, for the extra time delay caused by the Sun's curvature of nearby space amounts to only one ten-thousandths of a second.

Einstein's theory has now been verified by so many experiments that it is widely accepted as a brilliant contribution to our understanding of nature – begun by his attempts to account for Mercury's unexplained motion.

And now that we've considered both the expanding solar atmosphere and the Sun's curvature of nearby space, this brings us to the fullness of the space between the stars.

5.4 GAS AND DUST BETWEEN THE STARS

The celebrated American poet, T. S. Eliot, wrote, "O dark dark dark. They all go into the dark, the vacant interstellar spaces, the vacant into the vacant,"[61] but he was at least partially wrong. The interstellar spaces are decidedly full.

Light from the Sun and stars travels through space on its eternal trip to oblivion, and space is also filled with even greater amounts of the three-degree cosmic microwave background radiation. In every cubic centimeter of space in the Universe, there are 410 microwave photons, tiny bundles of radiation energy that originated about 14 billion years ago, when the first atoms formed and allowed the radiation produced in the Big Bang to travel freely.

And there is a rich variety of interstellar material, first suggested by emission nebulae in the immediate vicinity of very luminous stars. Their energetic starlight heats, ionizes and illuminates the rarefied surrounding material (Fig. 5.7). But only rare, hot luminous stars can ionize nearby hydrogen, and the vast majority of interstellar space therefore consists of much colder, invisible and unionized, or neutral, hydrogen atoms, detected by radio astronomers.

Interstellar Hydrogen

A Dutch graduate student, Hendrik C. "Henk" Van de Hulst, first proposed a method of detecting the radio emission of interstellar hydrogen atoms in 1945. He realized that a rare collision between two of the atoms could result in a change in the direction of the spin, or rotation, of the electron in one of the atoms. The atom is then in an unstable configuration, so its electron will soon flip back to its original spin direction. This releases a small amount of energy and produces radiation at a wavelength of 21 centimeters.

The remarkable story of the detection of the 21-centimeter line follows a paper trail of international scope. The prediction was published by Van de Hulst in 1945 in an obscure Dutch journal *Nederlands tijdschrift voor natuurkunde,* with an obtuse article title of "Radio Waves from Space: Origin of Radio Waves." The Soviet theorist

FIG. 5.7   Illuminating space The *Hubble Space Telescope* has resolved this glowing cloud of interstellar gas, known as NGC 604, which is in the nearby galaxy M33,
located about 2.7 million light-years away. Ultraviolet light from about 200 brilliant,
blue, relatively young stars ionizes the surrounding gas; the exited ions fluoresce and
produce the radiation seen in this image, which is 1,300 light-years across. The most massive stars exceed 120 times the mass of our Sun, with disk temperatures as hot as 40,000
kelvin. Billows of interstellar dust also surround and penetrate the nebula, hiding some of
the illuminated material from view. (Courtesy of NASA, AURA/STScI and the Hubble
Heritage Team.)

Iosif S. Shklovskii confirmed the prediction four years later, with greater detail, but in
the Russian language. At about this time, Harold I. "Doc" Ewen, a graduate student at
Harvard University, became interested in radio astronomy, and his advisor, Edward M.
Purcell, asked his wife, Beth, who was good at languages, to translate Shklovskii's paper
about the 21-centimeter line. After reading it Purcell encouraged Ewen to build a radio

receiver to search for the transition. They also had a copy of Van de Hulst's original 1945 work, translated from the Dutch.

Ewen constructed a radiometer to detect the hypothetical radiation using electronic components scavenged from other places, or brought with a $500 grant and $300 from Purcell's pocket. Much of the equipment was borrowed every Friday, and returned each Monday, from Harvard's Cyclotron Laboratory using a wheelbarrow. Accurate measurements of the expected hydrogen line wavelength, using terrestrial atomic hydrogen in a laboratory at Columbia University in 1950, enabled Ewen to tune his receiver to the precise wavelength of 21.106 centimeters. And at Purcell's suggestion, the receiver was switched between the wavelength of the expected signal and an adjacent one, with a difference that might contain the expected 21-centimeter line.

The receiver was connected to a simple horn antenna that was built of plywood, lined with copper sheeting, and mounted on a ledge just outside a window in Harvard's physics building. And by March 1951, Ewen had succeeded in detecting the 21-centimeter transition with the novel receiver, shortly past midnight when the Earth's rotation brought the plane of the Milky Way through the beam of the horn antenna.

As it turned out, Van de Hulst was in Cambridge, Massachusetts as a visiting professor at Harvard College Observatory, and the Australian radio astronomer Frank Kerr was visiting Harvard on a Fulbright grant. So Ewen and Purcell had them over, describing their discovery and urging them to have their people confirm the result.

At the meeting, the Harvard team learned for the first time that the Dutch group at Leiden had been actively trying to detect the radio transition for several years. So a description of the receiver was provided to Van de Hulst, leading to the conversion of the Dutch system, and in a gracious move, which would be unheard of in today's competitive scientific world, Purcell insisted that publication of their discovery be held up until the Dutch group confirmed it.

Even though the Australians were not actively pursuing a search for the hydrogen transition, and did not have a detection system in operation at the relevant wavelength, two groups assembled the necessary components soon after receiving word from Kerr, and repeated the detection about a month after the Dutch had. A coordinated report from all three centers was announced in the journal *Nature* in July 1951.

The detection of the 21-centimeter line revolutionized studies of the interstellar medium. The radio waves propagate right through the curtains of interstellar dust that hide most of our Galaxy from view at the optical wavelengths of visible light. And within a few years the Leiden and Australian groups were able to map out the intensities and velocities of the 21-centimeter emission in different directions, detecting the coiled, spiral shape of our Galaxy.

Stellar Bright and Cloudy Dark

The contrasting tones of light and dark accentuate our surroundings, helping us to see. They are what make sunrise, sunset, or a stormy day so beautiful, and gives Paris, the city of lights, its charm at any time of day or night. This interplay of light and shadow can also be seen in the Milky Way on any clear night, where the dark places create shadowy silhouettes against nearby bright stars (Fig. 5.8).

FIG. 5.8   Horsehead Nebula A long, bright strip of ionized hydrogen, designated IC 434, illuminates the dark, dusty Horsehead nebula, also known as NGC 2024. The bright circular object to the left of the Horsehead nebula is the star Zeta Orionis, which is the eastern star in the belt of the constellation Orion. A fainter, hotter star, Sigma Orionis, lies just above the Horsehead. Its excessive ultraviolet light ionizes IC 434. (Courtesy of David F. Malin, Royal Observatory, Edinburgh © 1980).

It was Edward Emerson Barnard who contributed most to our early awareness of these dark regions, at a time when most astronomers thought that they were an unfathomed emptiness, literally holes in the sky. His family was poor, and by the time he was nine years old Barnard had begun to work as an assistant in a photographic studio, learning techniques that would be invaluable in his later career in astronomy.

Barnard became an accomplished portrait photographer, and didn't begin serious astronomical work until the age of thirty, when he moved to the Lick Observatory on the isolated Mount Hamilton in California – where he systematically photographed the dark and bright regions of the Milky Way using wide-field portrait lenses. This work extended over thirty years, first at Lick and then at the Yerkes Observatory near Chicago, and culminated in two stunning catalogues of the regions, which he noncommittally called dark markings (Fig. 5.9). Their nebulous form and shape suggested to

FIG. 5.9 Dark clouds in Taurus Interstellar dust blocks the light of distant stars, marking the location of dark filaments that contain newborn stars. This photograph, published by Edward Emerson Barnard (1857–1923) in 1927, is 9.6 degrees in angular width, which corresponds to a linear width of 78 light-years at the distance of the Taurus clouds.

Barnard that the dark places were not empty, and in some cases they even seemed to interact with the bright regions that enfolded them.

Like Barnard, the German astronomer Maximilian Wolf at Heidelberg also recognized the importance of photographing the Milky Way with instruments having a large field of view. While Barnard was chiefly interested in the peculiar shapes of the dark regions, Wolf was concerned with measuring their distances and absorbing powers. By counting stars in an obscured and an adjacent unobscured region, Wolf was able to show in 1923 that the dark areas are obscuring clouds that absorb the light of distant stars. Because Wolf could not detect any substantial difference in the colors of stars that lie outside and behind the dark nebulae, he concluded that the dark regions must be composed of solid dust particles, whose scattering properties depend weakly on wavelength, rather than gas atoms, which scatter light much more effectively at shorter wavelengths.

Interstellar Dust

Since nothing could be seen that might absorb starlight outside of the dark clouds, most astronomers initially assumed that the rest of interstellar space was transparent.

But then, in 1930, Robert J. Trumpler of the Lick Observatory found persuasive evidence for unseen regions of obscuration throughout the plane of our Galaxy.

The Swiss-born astronomer was studying the properties of open star clusters, such as the Hyades and the Pleiades, located in the disk of our Galaxy, along the plane of the Milky Way. After more than a decade of investigation, he published a lengthy and comprehensive study of one hundred open clusters, discovering an inconsistency that convincingly demonstrated pervasive interstellar absorption of starlight.

Trumpler first used the colors of the brightest stars in each open cluster to infer their intrinsic luminosity, and then calculated the distance from their apparent and intrinsic luminosities. But when these distance estimates were combined with the measured angular diameters of the open clusters, he found that the linear diameters increase with distance, and the difference wasn't trivial. In whatever direction he looked, more remote clusters seemed to be about twice as large as the closer ones. Moreover, the effect was systematic, the further away a cluster was, the larger it appeared to be.

Concluding that this pervasive, systematic change in physical size was impossible, Trumpler instead assumed that all open clusters really have the same linear diameters or physical extent. This meant that the initial distance estimates had been overestimated due to the absorption of starlight by interstellar dust with an amount that increases with distance. The greater the distance, the more the absorption, making the remoter clusters look systematically fainter and even further away than they actually are.

But just what is this dust that is densely packed within dark clouds and distributed throughout interstellar space? Because the cosmic dust blocks and scatters light, the individual particles are about the same size as the wavelength of light. That is roughly four ten-millionths, or 4×10^{-7}, meters across, which is about the same size as the particles of cigarette smoke or the dust motes you see when looking back at a movie projector.

The dust is found in greatest concentrations within the dark interstellar clouds, which it refrigerates by blocking high-energy stellar radiation. It would otherwise heat the clouds and destroy any molecules that might be formed there. These molecules emit radio waves that can escape the dusty regions without attenuation, just as your radio station sends its signals through the cloudy, dusty air.

Molecular Cocktails in a Smoky Room

Soon after the discovery of interstellar atomic hydrogen at radio wavelengths, astronomers began to speculate about the possibility of detecting molecules with radio waves, which are sensitive to the coldest clouds of interstellar matter. But observations of any molecule first required accurate measurements of the wavelength of its spectral features in the terrestrial laboratory. Furthermore, radio telescopes with surfaces accurate to a few centimeters or better had to be constructed for receiving the short-wavelength emission of molecules, and new methods of spectral analysis had to be developed.

Not until 1953 did Charles H. Townes and his colleagues at Columbia University make the first precision laboratory measurements of the radio frequency transitions of the hydroxyl, OH, molecule, composed of an atom of oxygen, O, and an atom of hydrogen, H. And at an international symposium of radio astronomy in 1955, he presented laboratory measurements of the rotational transitions of molecules that might be detected, including carbon monoxide, water and OH.

By 1963, Alan Barrett and his colleagues at the Massachusetts Institute of Technology had built the sophisticated equipment needed to obtain the first observations of interstellar OH at 18.005 centimeters, the precise wavelength specified by Townes. The discovery was closely followed by intensive searches for OH, and eventually resulted in the realization that some of these sources are as small as stars, acting like cosmic masers – maser is an acronym for the microwave amplification by stimulated emission of radiation.

Townes was very familiar with masers, having constructed the first working terrestrial maser in a laboratory at Columbia in 1953, and sharing the 1964 Nobel Prize in Physics for his invention. In 1967 he moved to the University of California at Berkeley, to extend his previous work in microwave spectroscopy to the sky, and within a year Townes and his graduate students had discovered ammonia and water in interstellar space. This was soon followed by the detection of the embalming fluid formaldehyde in 1969, and carbon monoxide and hydrogen cyanide by other groups in 1970.

These discoveries triggered an avalanche of molecular searches in which groups of young radio astronomers armed with the latest laboratory measurements engaged in an extraordinarily competitive fight to be the first to detect the next interstellar molecule. The net result has been the discovery of a pharmaceutical array of hundreds of interstellar molecules, including complex organic molecules such as ethyl alcohol, or ethanol, the substance that gives beer, wine and liquor their intoxicating power.

The molecules reside within dense, dark and dusty clouds that typically span up to 120 light-years and harbor a total mass of up to a million times the mass of the Sun, mainly in the form of molecular hydrogen with a density of up to 300 molecules per cubic centimeter. The dust serves mainly to block out the harsh radiation in space, and enable chemical reactions to form complex, delicate molecules from the atomic constituents of the interstellar gas. These giant molecular clouds are the dominant component of the interstellar medium, with a combined mass of about 10 billion Suns. Yet, they are black and exceedingly cold, with temperatures of about 10 kelvin, radiating almost exclusively in the microwave region of the electromagnetic spectrum.

Under the right circumstances, a giant, massive molecular cloud collapses under its own weight, eventually forming up to a million stars. In fact, some stars have already formed in these clouds. They are the present-day incubators of newborn stars.

5.5 DARK MATTER – EVIDENCE OF THINGS UNSEEN

Since childhood, we've learned to be afraid of the dark. The velvety-black night can be a bit scary to adults as well, especially since no one knows exactly what might be out there. As the old refrain goes, "only the shadow knows." Or perhaps "beware the dark side" is more appropriate.

Dark has been associated with evil, and light with good, since at least the time of Zarathustra about 1300 BC. Dante's divine journey took him from the dark forest to the radiance of paradise. And today we have Darth Vader in *Star Wars,* and black holes, which are about as dark as you can get.

Darkness pervades the Cosmos, occupying most of its space. Look down at the Earth from space, for example, and you only find isolated fires or city lights that illuminate the darkness of the night, and you can't see all the hidden things related to our presence (Fig. 5.10). Most of the Universe is similarly invisible to us, giving off neither light nor radio waves to let us know it is there. Luminous stars and galaxies, the things that shine, are mere white caps on a dark invisible sea, controlled by influences beyond our vision. Shakespeare seems to have alluded to it with "Now entertain conjecture of a time, when creeping murmur and the poring dark fills the wide vessel of the Universe."[62]

FIG. 5.10   Europe's lights This striking image from space shows Europe at night. Lights from major cities illuminate tiny spots within the countries and help delineate their coastlines. But most of the land, and the surrounding oceans, are dark and invisible at night. (Data courtesy of NASA, NOAA, Marc Imhoff and Christopher Elvidge, image by Craig Mayhew and Robert Simmon, NASA GSFC.)

The astronomer Agnes Clerke noticed its possible cosmic implications a century ago, when the entire Universe was thought to consist only of stars. In 1903 she wrote of dark stars, whose presence might be inferred from their gravitational effects on the motions of visible stars, and stated that "unseen bodies may, for ought we can tell, predominate in mass over the sum total of those that shine. They supply possibly the chief part of the motive power of the Universe."[63]

Although the elusive dark matter can't be seen in visible light, we now know that it affects things that we can see. Its gravity keeps clusters of galaxies from flying apart, and holds rapid, swirling stars and gas in at the edges of galaxies. By observing these motions, astronomers can infer the presence of dark matter, somewhat like watching leaves carried by an invisible stream. They show us that the radiant stars and galaxies provide an incomplete description, telling us no more about the Universe than city lights tell us about a nation.

Invisible Matter in Clusters of Galaxies

Evidence that the Universe might contain substantial amounts of unseen matter has been around for a long time, ever since the 1930s when the Swiss astronomer Fritz Zwicky, at the California Institute of Technology, determined the total mass of the Coma cluster of galaxies. He measured the amount of mass required to keep the cluster stable, assuming that the motions of its constituent galaxies are balanced by the gravitational pull of their combined mass, and found that the total mass of the Coma cluster is abut ten times the sum of the masses of the visible galaxies it contains. If this hidden mass were not there, the cluster would be flying apart.

In an extraordinarily prescient paper, entitled "On the Masses of Nebulae and of Clusters of Nebulae," published in the *Astrophysical Journal* in 1937, Zwicky concluded that there must be substantial amounts of invisible intergalactic gas in clusters of galaxies, or else they would be dynamically unstable. In a German language article discussing many of the same topics four years earlier, Zwicky introduced the term *dunkle materie,* or "dark matter" for the invisible stuff, and concluded "if this is confirmed, we would arrive at the astonishing conclusion that dark matter is present with a greater density than luminous matter."[64]

In his 1937 paper, Zwicky also proposed that the formidable gravity of dark matter in clusters of galaxies would act as a powerful lens, diverting and focusing the light of more distant galaxies. As he suggested, the effect can be used to estimate the mass of the lensing material. But it wasn't until the early 1980s that observations of multiple quasar images, produced by a lensing foreground galaxy, put the concept of cosmic gravitational lenses on a firm observational footing.

This was soon followed by the detection of the distorted images of distant galaxies seen through compact, massive clusters of galaxies. As discovered by Roger Lynds and Vahé Petrosian in the late 1980s, the entire cluster spreads the light of a remote background galaxy into an array of luminous arcs, providing information about extended, massive concentrations of dark matter within the cluster of galaxies. *Hubble Space Telescope* images of rich clusters of galaxies have subsequently revealed the highly stretched, distorted and magnified images of faint galaxies lying far behind them (Fig. 5.11). They

FIG. 5.11　Cluster of galaxies and gravitational lens A *Hubble Space Telescope* image of a rich cluster of galaxies designated Abell 2218, number 2218 in George Abell's (1927–1983) catalogue. A typical rich cluster contains hundreds and even thousands of galaxies, each composed of hundreds of billions of stars and possibly up to ten times more mass within invisible dark matter. The powerful gravitation of the compact, massive cluster deflects light rays passing though it, acting like a lens that magnifies, brightens and distorts images of objects that lie far beyond the cluster. It has spread the light of very distant galaxies into an array of faint arcs. (Courtesy of NASA.)

have additionally shown that galaxy clusters can act like a zoom lens, magnifying distant galaxies too faint to be seen and bringing them into view.

Hot, X-ray emitting gas has also been found permeating the space between galaxies in massive clusters. Any hydrogen atoms immersed in these clusters and moving at a similar speed to the galaxies would have to be very hot, with a temperature of about 100 million kelvin. At this temperature, the gas is an intense emitter of X-rays.

Although most of the observable mass of the clusters of galaxies is in the form of hot, X-ray emitting gas, outweighing all the visible-light stars within all the galaxies by a factor of about seven, both the stars and the intergalactic X-ray emitting gas constitute only about 15 percent of the total mass of the galaxy clusters. The remaining 85 percent of the gravitating material is some kind of mysterious dark matter, emitting no detectable radio, visible-light or X-ray radiation, and no one knows what it is.

Keeping the Spin on Spiral Galaxies

Look down at a modern city from an airplane at night, and the downtown area might be brightly lit. But that's not where many of the people are. They're out sleeping in the suburbs, which emit relatively little light.

A similar thing happens in the remote parts of spiral galaxies, where there is a lot of dark matter and few stars. The hidden matter exerts a gravitational force, which betrays its presence.

By far the brightest part of a spiral galaxy is its nucleus or central bulge. So astronomers naturally thought that the center was the most massive region, which gravitationally controlled the motions of the stars outside it. In this case, the stars would all revolve around the massive center at speeds that decrease with distance, just as the more distant planets move with slower speeds around the central Sun, which is overwhelmingly the most massive part of the Solar System. But the rotation speeds of stars and gas situated at the outer visible edges of spiral galaxies stay level as far as astronomers can measure. And that means the spiral galaxies don't end where their light does. At still greater distances, well outside the boundary of the visible galaxy, there are appreciable amounts of dark, unseen matter. It keeps the fast-spinning visible material connected to the galaxy.

This totally unexpected result was discovered by measuring the orbital, or rotational, speeds of gas or stars far from the centers of nearby spiral galaxies. Such motions are inferred from the Doppler shifts of spectral lines of bright stars or emission nebulae, seen at optical wavelengths, or interstellar hydrogen gas detected at radio wavelengths near 21 centimeters. Vera C. Rubin, at the Carnegie Institution of Washington, D.C., was among the pioneers, examining the motions of stars across the disks of spiral galaxies. During the 1970s and 1980s, Rubin and her colleagues showed that the outlying stars in the visible disks of Andromeda, or M31, and other nearby spiral galaxies move just as fast as the stars near their central bulge. Substantial amounts of unseen, non-luminous matter are required to keep the stars moving out of the galaxy's control, like a discus thrown out of a whirling athlete's hands.

During the same period, several radio astronomers showed that remote clouds of hydrogen atoms are spinning about the centers of spiral galaxies unexpectedly fast. The 21-centimeter rotation velocity remains constant with increasing distance from the center of spiral galaxies, well beyond the visible stars (Fig. 5.12). Since the rotation curves are flat out to at least twice the extent of the visible galaxy, astronomers estimate that each spiral galaxy is immersed within, and enveloped by, a halo of dark matter at least 10 times bigger and heavier than its visible component, extending as far as a million light-years. If this were not so, the galaxies could not retain their high-speed outer parts; they could not stay together and would fly apart.

There is more than meets the eye in our Galaxy as well. The mass of its visible inner regions, inferred from the orbital motion of the Sun about the galactic center, is 100 billion Suns. But the rapid motions of dwarf satellite galaxies, which revolve about our Milky Way at distances of up to a million light-years, indicate that a great reservoir of unseen matter also envelops our Galaxy. In order to hold onto these companions, our Galaxy must have a total mass of roughly two trillion, or 2×10^{12}, times the mass of the Sun, and roughly ten times the mass of its visible stars.

In other words, there seems to be at least ten times more dark material than luminous stars hidden within the vast unseen places that envelop our Galaxy. Most of this mass does not lie within the visible plane of the Milky Way, but beyond it in a dark halo, which outweighs the rest of the Galaxy by a factor of ten and extends out to a million light-years from the galactic center.

At Princeton University, Jeremiah P. Ostriker and P. James E. Peebles additionally concluded, in 1973, that a dark, invisible halo is required to constrain and mold other spiral galaxies from outside, helping them to retain their distinctive shape. The flat spiral disks cannot remain stable unless they are surrounded by a huge, roughly spherical

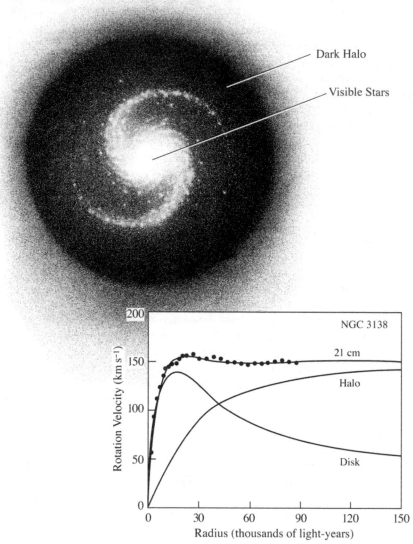

FIG. 5.12   Dark matter envelops a spiral galaxy The rotation velocity of the spinning spiral galaxy NGC 3198 plotted as a function of radius from its center *(bottom)*.
The observed neutral hydrogen 21-centimeter data are attributed to an optically luminous
disk, containing all the visible stars, and a dark halo that contributes most of the mass
at distant regions from the center. The visible stars and surrounding halo of dark matter
are illustrated in a hypothetical drawing of a spiral galaxy seen from above *(top)*. The
fact that the rotational speed of the cool hydrogen gas remains high even at the largest
distances indicates that the outermost gas must be constrained and held in by the gravitational pull of dark matter far outside the visible part of the galaxy. [Adapted from T. S.
Van Albada and colleagues, "Distribution of dark matter in the spiral galaxy NGC 3198,"
Astrophysical Journal **295,** 305 (1985)].

halo of invisible matter. As with the Milky Way Galaxy, the unseen haloes of these spiral galaxies extend out to about a million light-years, each containing a mass of a trillion, or 10^{12}, times that of the Sun. The visible stars are sandwiched within the dark halo, like a hamburger in its bun. In answer to the question "where's the beef?" they might reply "it's all in the bun, where the dark matter resides." About 90 percent of the mass of galaxies has to reside in a much larger invisible halo of unknown composition.

Astronomers also speculate that the invisible blackness, which now enfolds the luminous galaxies, might explain where the galaxies came from, and why they have the sizes and shapes that we observe. After all, if dark matter is around now, it might have been around back at the beginning of time, outweighing everything else and providing the dominant gravity needed to pull the galaxies together.

Inducing Galaxy Birth

Dark matter offered a possible solution to a problem that has troubled astronomers for decades. The unexplained paradox is that ordinary matter, collectively known as baryons, could not have coalesced quickly enough after the Big Bang to make the galaxies that we now observe.

When left alone, a spread out distribution of matter will coalesce about small, initial concentrations, as a result of their gravitational pull on surrounding material, and any slight perturbations in otherwise uniform matter will grow and eventually collapse. But the expansion of the Universe wrecks havoc with this scenario. Because the expansion pulls the primeval material apart as soon as it starts to come together, tiny density fluctuations grow too slowly to create the structures that are seen today. There just isn't enough time in the entire past history of the expanding Universe for gravity to gather chance density fluctuations into clusters and superclusters of galaxies.

The extra gravitational pull of some sort of dark matter seemed to be required to clump and shape the expanding Universe into the observed concentrations of luminous galaxies. In other words, gravity can only outpace and resist the overall expansion of the Universe if dark matter loads the dice of the creation game, ruling the Cosmos even before perfectly ordinary visible matter began to interact gravitationally.

But just what is this mysterious invisible matter that pulled the Universe into its present-day shape? Astronomers and physicists have been arguing about it for decades, and there still isn't a consensus. And in the meantime, other scientists have even been speculating that the entire observable Universe originated from vacuous, empty space, with everything coming out of nothing.

5.6 MAKING EVERYTHING FROM NOTHING

Inflation

Shortly after the discovery that the galaxies are in flight from each other, and from us, it was realized that their expansion must have originated from a highly compressed and extremely hot state. You can visualize this starting point by putting the observed expansion in reverse, like running a movie backwards, pushing the galaxies back close together.

One of the first persons to investigate the physics of this early state, when all the matter was compressed to an enormous mass density, was the Belgian astrophysicist and Catholic priest, Georges Lemaître. He counted back until he reached time zero, his "day without a yesterday," when time began and the expansion of the Universe was jump-started from an incredibly dense "primeval atom" or "cosmic egg," sending its contents out in all directions.

Lemaître tried to connect this beginning of the observed world to quantum theory, which deals with the smallest possible components of a system, speculating in 1931 that the beginning of the Cosmos might have occurred as a single quantum "a little before the beginning of space and time."[65] So, Lemaître was advocating a singular beginning of everything – matter, radiation, energy, space and time – a very long time ago when the Universe was very small.

Quantum theory matured over succeeding decades, and several theoreticians eventually attempted to extend it to the origin of the Cosmos. In a paper entitled "Is the Universe a Vacuum Fluctuation", published in 1973, the American physicist Edward Tryon proposed that our Universe originated as a quantum fluctuation in a vacuum, providing the first scientific mechanism by which the Universe might spontaneously have arisen from "nothing". And the Russian physicist Alexei A. Starobinsky showed, in 1979, that quantum gravity might imply that the youthful Universe became exponentially large, expanding at a rapid, accelerating pace. The Dutch astronomer Willem de Sitter anticipated such an exponential growth in 1917, when he discovered that Einstein's *General Theory of Relativity* could imply a rapid expansion to huge dimensions even without any matter around.

Alan H. Guth, a relatively unknown, 34-year-old American particle physicist, also speculated in the early 1980s that the young Universe expanded with pronounced haste. According to Guth, the Universe went through a period of accelerated expansion during the first miniscule moments of its existence, boosting the nascent Universe in size from almost nothing to literally cosmic proportions.

When the Universe was just 10^{-35} seconds, or 10 trillion trillion trillionths of a second, old, space was supposed to balloon outward at a faster and faster rate, increasing in size by a factor of 10^{25}, or a ten million billion billion. Within an infinitesimal fraction of a second, a vacuous, sub-atomic bubble became larger than anything we could ever see, even with our most powerful telescopes. Guth called this growth spurt inflation, like inflating a balloon with hot air rather than in the economic sense of rising consumer prices.

Guth's model was based upon a Grand Unified Theory, or GUT for short, which was proposed in 1974 to combine the weak, strong and electromagnetic interactions of elementary particles into a single force. The three interactions were supposed to appear as a symmetric unity in the very hot, immediate aftermath of the Big Bang, but when the early Universe expanded and cooled, they crystallized out of their common union, breaking their symmetry. This transition released a tremendous amount of energy all at once, inflating the tiny, embryonic bubble of practically empty space and stretching it to beyond astronomical size.

Symmetry is a way of saying that there are no observable differences when looking in various directions, at least about an axis of symmetry. And metaphorically, the broken symmetry is similar to the transition from liquid to freezing water. If you were submerged in a tank of liquid water, it would look the same in every direction. But this symmetry is broken when the water freezes into ice, forming cracks and an asymmetric crystalline structure while giving off a surprising amount of heat. As anyone who has left a Coke can in the freezer of a refrigerator knows, the freezing water also expands.

The Universe was still so young when inflation occurred that only space was supposed to exist, without any matter or radiation. But by the time inflation came to an end, at a time far less than one second in the Universe's history, the inflation had converted empty space into material particles and hot radiation, expending its inflationary strength like a hurricane that is traveling inland. So the energy released during the effective crystallization of space could have been converted into all the matter and energy of our observable Universe, effectively making everything out of nothing, like a magician who can pluck a card, a bird, or a rabbit out of apparently empty space.

As Guth subsequently remarked, "conceivably, *everything* can be created from nothing. And 'everything' might include a lot more than what we can see. In the context of inflationary cosmology, it is fair to say that the Universe is the ultimate free lunch."[66]

Over two thousand years ago, the great Roman poet and philosopher Titus Lucretius Carus wrote, in his epic poem *On the Nature of Things*, "nothing can be created out of nothing." But here was Guth, proclaiming that he knew how to make everything out of nothing.

Unfortunately, there was no way to gracefully stop the hypothetical inflation once it started, returning the Universe to the gradual expansion that we observe today. Since the expansion was occurring in empty space, it couldn't tell what it was expanding in reference to, and therefore didn't know when or how to stop. It seemed to be out of control, like a runaway car, and the only way to end the inflation involved disastrous collisions that are in disagreement with the observed Universe. Guth therefore renounced his model, after investigating its implications for about a year. And no one now believes his original idea, in which inflation was supposed to be related to a phase transition occurring at the breaking of a hypothetical grand-unification symmetry.

But the Russian astrophysicist Andrei Linde solved some of the problematic aspects of the early versions of inflation theory in 1983 when he proposed chaotic inflation, which is also rooted in the theory of elementary particles. In particular, the so-called scalar fields, which were introduced to explain particle interactions, can also introduce instability into the nothing of the very early Universe, causing inflation. And because these primeval fields can take on arbitrary values, Linde called this scenario chaotic inflation.

According to this theory, space contains scalar-field energy, even when there isn't any matter. The persistence of this energy can lead to a stage of extremely rapid expansion, or inflation, of the Universe, and when the energy decreses the inflation we know about ends.

The clever aspect of Guth's original inflation hypothesis, which accounts for inflation's persistent popularity, was its ability to possibly resolve two cosmic conundrums, known as the flatness and horizon problems. They are related to the apparent lack of curvature, or flatness, of space, and to the smooth, homogeneous structure of the cosmic microwave background radiation. The swift, fierce acceleration inflated these vexing problems away during a fleeting moment at the dawn of time, ironing the observable Universe out flat and stretching out any irregularities in the radiation.

Flatness is a two-dimensional way of picturing a three-dimensional space that is described by Euclidean geometry, in which parallel lines remain parallel and never meet. In curved space, which is described by non-Euclidean geometry, parallel lines can either diverge and widen, or converge to a meeting point.

If there is no cosmic repulsion, and the infamous cosmological constant is zeroed out, then the curvature of the Universe is determined by the mutual gravitational attraction of matter, which fights against the swelling of space. If the density of all the mass in the Universe is significantly higher than the critical value required to gravitationally rein in the expansion, then the Universe has positive curvature and will be closed in the future; and if the density is noticeably lower than the critical value, then the curvature is negative with a wide-open, ever-expanding Universe.

The inflation theory proposes that the present mass density is on the borderline, at the critical value. It assumes that the Universe is not curved, forever balanced at the knife-edge of closure, while also supposing that the cosmological constant does not exist.

Any deviation from this perfect balance at the beginning of the expansion would have been magnified over time, causing today's mass density to depart from the critical value by an enormous factor. It's analogous to the way a small difference in the way a baseball is hit by a bat can produce a big difference in the baseball's trajectory, as either a foul ball or a homerun. Yet, observations indicate that the present mass density of galaxies is fairly close to the critical value, perhaps within a factor of ten of it. The expanding Universe therefore might consist of "flat" space without curvature, described by Euclidean geometry.

Inflation provides a mechanism for straightening out the geometry of the Universe near the very beginning – by bloating it to colossal size and pushing most of it out of sight. And a corollary of this explanation is that the observable Universe might be just an insignificant part of the entire Cosmos, perhaps not even a representative part.

The "horizon" problem relates to the extraordinary smoothness of the three-degree cosmic microwave background radiation, whose temperature fluctuations are less than one part in one hundred thousand. This means that two regions on opposite sides of the microwave sky have precisely the same temperature, and look the same even though they lie beyond the visible horizon of either one of them. They are so far apart that there has not been time since the Universe began for light and any other signal or physical influence to pass between them and synchronize their properties, bringing them into uniformity.

Inflation explains this paradox by stretching the Universe out, accelerating the expansion to unimaginable speeds even faster than light. The Universe became so large so fast that we can now see just a tiny fraction of it, which appears smooth and uniform.

The inflationary expansion of the Universe should have stretched out, and erased, all traces of any characteristic angular scales in the cosmic microwave background radiation, creating a flat power spectrum of angular sizes that has no distinguished features. But the observed spectrum (Section 4.7) lacks power in the largest angular sizes, which is inconsistent with inflation models, and the ripples found in the power spectrum at other angular sizes require the gravitational pull of invisible dark matter.

For most astronomers, the uncomfortable aspect of the inflation idea was that the mass of all the ordinary baryonic matter in the Universe, in both visible and invisible form, is too small, and its gravitation too weak, to fight off the expansion of the Universe. Proponents of the theory nevertheless argued that the required mass might reside in some sort of exotic elementary particles that no one had ever seen before. Physicists have been hunting for them for decades, but no one has yet caught a glimpse of the imaginary stuff (Focus 5.2).

After decades of fruitless search and speculation, none of the elusive, dark-matter particles have been found. There is no evidence that any of them exist, and even if there were it would be extremely indirect. When asked about the current situation, one Professor of High-Energy Physics said: "It's all hype and pure speculation. But don't ask the dark-matter particle experts about it, they're all members of the brotherhood and won't say a word."

FOCUS 5.2

Ghosts in the Attic

Inflation advocates and particle physicists have speculated that the Universe contains large numbers of non-baryonic particles, produced in the Big Bang and still hanging around. Billions of these particles could be whizzing harmlessly through your body every second, the ghostly shadows of creation in the Big Bang.

The leading dark-matter candidate early on was the hot, fast-moving neutrino, the only non-baryonic particle known to exist. Enormous numbers of neutrinos were almost certainly produced in the Big Bang, and they are still around, greatly outnumbering all the atoms in existence. Calculations indicate that there are about one billion of these neutrinos for every hydrogen atom in the Cosmos. If each one of these neutrinos had even a tiny mass, their combined weight could exceed that of all ordinary baryonic matter. However, in 1995 the subterranean Superkamiokande neutrino detector showed that even the heaviest kind of cosmic neutrino has to be at least five billion times less massive than a proton, which meant that the combined mass of all

the neutrinos in the Universe only contributes between 0.1 and 5 percent of the critical density needed to bring the Universe to the brink of closure.

With the demise of the neutrinos, came the rise of the cold, slow-moving WIMPS, an acronym for Weakly Interacting Massive Particles. Compared with the lightweight neutrino, which is much less massive than a proton, the WIMPS could be the heavyweight champions of the Universe, each one weighing in with a mass of 10 to 100 times that of a proton. Their combined gravitational muscle could rule the Cosmos.

For a while, the WIMPS were declared winners, triumphing over all the other imaginary particles. But the WIMP victory was a premature conclusion. If they are out there no one has been able to find them so far. After decades of fruitless search and speculation, none of the elusive particles have been found.

By 1998, John Maddox, the influential, former editor of the journal *Nature* was able to declare that this continuing search for enough missing mass to close the Universe, amounting to perhaps 80 percent of the total mass, "is the modern equivalent of the fruitless search for the luminiferous aether a century ago."[67]

So the hypothetical dark, sub-atomic entities are starting to resemble something conjured up by Sauron in Tolkein's *Lord of the Rings,* or by Darth Vader in the movie *Star Wars.* And it is likely that the hypothetical non-baryonic particles will recede into the dustbin of history. But in the meantime, scientists have discovered another hypothetical way of making everything out of nothing, and it is called string theory.

Strings

Like inflation, string theory grew out of theoretical models for sub-atomic elementary particles and supposedly empty space. When examined on smaller and smaller sizes or shorter and shorter time scales, space gets increasingly agitated, like crowding too many people on a small dance floor or trapping a claustrophobic person in a stalled elevator. Nothing likes to get cornered, squeezed into a place it can't get out of, and that includes space.

The reason for all this microscopic turbulence is the Uncertainty Principle, discovered in 1927 by the German physicist Werner Heisenberg. It states that there are features of the Universe, such as the position and velocity of a sub-atomic particle, which cannot be simultaneously known with perfect exactitude. As Heisenberg noticed, one consequence is that what we see depends on how we look at it, or that "what we observe is not nature in itself, but nature exposed to our method of questioning."[68]

As a result of the Uncertainty Principle, even empty space can't stand getting closed within too small an increment, and there is a violent change in its condition on microscopic scales. This frenzy of vacuous space is supposed to produce the hot, tiny, short-lived strings, which might explain the forces of Nature at incredibly high energies in exceedingly small spaces and over very short time scales. According to the quantum string "magic," the agitation makes strings vibrate. But of course you can't see the vibrations, with even the most powerful instruments, and we never will; they appear only in certain mathematical equations that almost no one understands.

The Italian scientist Gabriele Veneziano proposed a mathematical foundation for string theory in 1968, as an explanation for nuclear particles and their interaction. By 1970, three other particle physicists realized that his equations describe the quantum mechanics of elementary particles, interpreted as little, vibrating, one-dimensional strings. And although the original string model was eventually replaced by a more convincing theory for the strong nuclear force, it was subsequently realized that string theory might have wider implications, perhaps unifying quantum mechanics, particle physics and gravity.

In a more evolved form, known since 1984 as the superstring theory, the revolutionary concept became a theory of everything, combining the laws of the large, smooth and curved, as expressed in the *General Theory of Relativity,* with the laws of the small, irregular, uncertain and random, described by *Quantum Mechanics.* All the things that physicists used to call particles and forces were replaced by just one basic ingredient – resonating, looping strings with different vibration patterns.

The oscillating strings, each an invisible loop of energy, were supposed to undergo quantum fluctuations or jitter, like some young punk hyped up on too much of the wrong kind of drugs, establishing the way things are. The electrons, for example, vibrate one way, and each force of Nature vibrates another way.

String theory accomplishes its unifying magic by introducing up to eight new dimensions hidden within the fabric of space. Elementary particles, like electrons, are confined to the three dimensions that describe the ordinary world we are used to. Our bodies are also trapped in the three spatial dimensions, but this space is supposed to be an illusion. Gravity, for example, may be able to escape into the extra dimensions, which are supposedly out there, but you can't see them, not ever.

At the beginning, in the Big Bang, all the spatial dimensions of string theory were curled up to their smallest possible extent, rolled up like coiled springs waiting to be released and form the world, or like a host of tiny spiders ready to launch and unreel the stringy filaments of their webs. And the vibrating strings might have themselves been produced way before the hypothetical inflation, at even smaller, hotter and denser times in the past, just 10^{-43}, or ten million trillion trillion trillionth of a second after the Big Bang.

Well, string theory is quite a comprehensive speculation. A host of wiggling little bits of invisible strings, which occupy unseen places and extra dimensions, are supposed to appear out of nothing and account for everything. Some physicists find it beautiful and elegant. It is elegant because the strings explain everything. But it is also very complicated. Even the eminent elementary-particle physicist Sheldon Glashow declared that he couldn't understand it.[69]

Moreover, string theory is bad science, at least in its present form. After more than thirty years of speculation, there is no direct experimental evidence for strings, and the theory does not make even one experimental prediction that can definitely prove it right or wrong. But it is that sort of predictive capability and observational verification that separates science from philosophy or religion.

At least one well known person seems to have gotten fed up with all the hype – the American movie actor, comedian and director Woody Allen who satirized it in a description of an interoffice romance in *The New Yorker* magazine, writing that he could feel his strings vibrating, but didn't know how to act since he couldn't determine an exact position and velocity and might cause a devastating rupture in space-time.[70]

Champions of string theory nevertheless claim that they have accomplished the very goal that eluded Einstein, who dreamed of a unified theory describing all the forces of Nature. Try as he would, Einstein never achieved this goal, but the young Turks are convinced that they have found the way, discovered the true theory of everything. One of them, the American particle physicist Brian Greene, has poetically described strings that "endlessly twist and vibrate, rhythmically beating out the laws of the Cosmos", with the caveat that "the mathematics of string theory is so complicated that, to date, no one even knows the exact equations of the theory."[71]

And to get back to the reality of the observable Universe, astronomers have shown that dark matter might no be so important after all. That's because matter isn't everything; there is energy as well. And since energy is equivalent to mass, invisible energy can take the place of unseen mass.

5.7 DARK ENERGY

Using supernovae as their beacons, astronomers have looked halfway across the Universe to measure the history of the cosmic expansion, clocking how much it has changed since the light was emitted from those distant exploding stars. And contrary to expectations, the supernova observations suggest that the expansion is speeding up as time goes on, rather than slowing down. Some pervasive, unknown repulsive force is apparently counteracting the combined, mutual gravitational attraction of all the matter in the Universe, pushing us faster and faster into oblivion.

Having no clue about the cause of this mysterious acceleration, astronomers have given it a name – dark energy. The enigmatic dark energy is supposed to spring out of nothing, from empty space itself. Moreover, the destiny of the Universe is no longer supposed to be determined by mass alone, and any geometry is possible, depending on the amount of dark energy hidden within empty space.

Measuring Cosmic Expansion with Exploding Stars

In the late 1980s, a group led by the physicist Saul Perlmutter at the Lawrence Berkeley National Laboratory began using a telescope to repeatedly observe large sections of the sky containing thousands of galaxies. A sensitive CCD detector and a computer were used to subtract images from the same region taken a few weeks apart. Anything that remained after the subtraction was a new source of light, present in a recent image and not in a previous one. It could be a type Ia supernova, which was confirmed by subsequent observations of its visible-light spectra taken with the Keck telescope in Hawaii or from the *Hubble Space Telescope*. The light curve of the distant supernova, a graph of its brightness over time, was additionally used to confirm its identity and sharpen the determination of its peak brightness.

Perlmutter's *Supernova Cosmology Project* faced some stiff competition by a rival group, dubbed the *High-Z Supernova Search*, formed in 1996. The "High-Z" denotes large redshifts and therefore supernovae in extremely distant galaxies. This second group, led by Brian P. Schmidt of the Australian National University, used similar techniques, whose feasibility had been demonstrated by Perlmutter, but it consisted of astronomers who had a lot of previous experience in observing supernovae, including their colors, spectra and light curves. It included Adam G. Riess at the Space Telescope Science Institute and Robert P. Kirshner of the Harvard-Smithsonian Center for Astrophysics.

The competition was basically a healthy thing, improving the quality of work by the rival teams. They both planned to measure the distances and velocities of supernovae far and near, determining how fast the Universe was expanding early and late in its history and how much the expansion might have slowed down as time went on. This was supposed to establish the total mass of the Universe, whose gravitational pull gradually slows the expansion.

In 1998 both groups startled the astronomical community by reporting that the Universe is speeding up as it flies apart rather than slowing down as expected. Their observations suggest that the expansion rate is significantly greater now than it was billions of years ago, when the distant supernovae emitted their light. So the Universe might not be slowing down after all, and some mysterious repulsive force seems to be counteracting gravity's restraining influence, accelerating the cosmic expansion.

The discovery of dark energy did away with the need for overwhelming amounts of dark matter to keep the Universe poised on the edge of future collapse, but never quite pulling it there. No more than one quarter of the critical mass density is now imagined to reside in mass of any kind, and astronomer's observations of such a low-density Universe are now widely accepted. Something even darker has taken over the Universe, perhaps providing the missing 75 percent required to keep the Universe at the brink of closure with "flat" space that is described by Euclidean geometry. And it is attributed to the outward pushing force of a hypothetical dark energy.

The trouble is, nobody understands the mysterious something that permeates space and eventually overwhelms the gravitational self-attraction of the entire material Universe. But it has been given some names, like the cosmological constant and quintessence, which are also poorly understood.

The Cosmological Constant

One approach to dark energy has been to resurrect the old concept of a cosmological constant, the anti-gravity force introduced and retracted by Albert Einstein more than half a century ago. In 1915 – 16 he invented his *General Theory of Relativity*, in which matter and energy distort or curve the geometry of space and time, producing the phenomenon called gravity. But when Einstein applied his equations to the Cosmos, most scientists thought the celestial Universe was eternal and unchanging, consisting solely of the fixed stars in the Milky Way. The unrelenting, universal gravitation of the matter in this static, or unmoving, Universe would exhibit a fatal attraction, eventually causing the collapse of the entire Cosmos under the combined weight of everything in it.

In 1917 Einstein therefore inserted a mathematical fix, the cosmological constant, into his relativity equations. The extra term represented the repulsive force of an unknown and undetected form of energy that permeated space and exerted a sort of outward pressure that opposed gravity. And Einstein adjusted the value of his cosmological constant so that its anti-gravity force would exactly counterbalance the gravitational attraction of matter, keeping the Universe from collapsing.

In a little more than a decade, Edwin Hubble had discovered that the galaxies are moving way from us in a cosmic expansion, so the Universe wasn't static after all. After visiting Hubble in 1931, Einstein abandoned the cosmological constant, and stated that the *ad hoc* term was greatly detrimental to the formal beauty of his theory. He instead collaborated with the Dutch astronomer Willem de Sitter in 1932 to propose an expanding Universe in ordinary uncurved Euclidean space, using the equations of the *General Theory of Relativity* without the cosmological constant.

But the artifice stubbornly refused to die, and has been repeatedly invoked whenever cosmologists have had trouble reconciling their theories with observations. All they had to do was revive the term, stick it in the relevant equations, and adjust its value; then all the headaches would go away, just like a good aspirin.

The cosmological constant was invoked for the second time in the 1930s when it was thought that the Earth might be younger than the expanding Universe. The great English astronomer Arthur Eddington regarded it as a fundamental constant of nature, using the term to obtain an age for the expanding Universe much larger than that implied from a backwards extrapolation of the estimated rate of expansion. As he wrote in 1933, a satisfactory theory would make "the beginning not too *unaesthetically abrupt*,"[72] with an initial static, or non-expanding, condition kept in balance by the cosmological constant.

At about the same time, Georges Lemaître constructed a model Universe with two periods of accelerated expansion, one at the beginning and one later, and a more gentle coasting period in between (Fig. 5.13). He proclaimed in 1931 that "the expansion thus took place in three phases: a first period of rapid expansion . . . a period of slowing-up, followed by a third period of accelerated expansion. It is doubtless in the third period we find ourselves today."[73]

This interpretation is somewhat similar to some modern explanations, beginning with rapid expansion, now called inflation, and with a currently accelerated expansion that might invoke the cosmological constant.

Although the expansion age of the Universe was eventually shown to be much older than the Earth, as the result of improved observations, the cosmological constant was revived in the 1990s to resolve controversial suggestions that the expanding Universe might be younger than its oldest stars. This time around, the resurrection of the cosmological constant not only removed this possible age crisis; it also solved the discrepancy between the observed amount of matter in the Universe and the amount needed to give space a Euclidean geometry and reconcile observations with the inflation theory.

FIG. 5.13   Models of the expanding Universe Schematic representation of various cosmological models showing the size, or scale factor R (t), of the expanding Universe as a function of time, t. The approximate age since the expansion began is given by $1/H_0$ where H_0 is Hubble's constant. Since the expansion age was once thought to be smaller than the age of the oldest rocks on Earth, Georges Lemaître (1894–1966) and Arthur Eddington (1822–1944) independently used a cosmical repulsion term with Einstein's *General Theory of Relativity* to

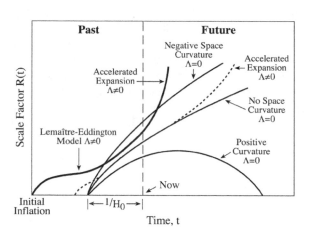

permit an adjustable age for the Universe. The adjustable term is symbolized by a non-zero value for the cosmological constant Λ. In this interpretation (*thick solid line*), the Universe began with expansion against gravity, followed by an essentially non-moving stagnation in which gravitation and cosmical repulsion were nearly in balance. This coasting period was then followed by an accelerated expansion driven by the cosmical repulsion. Three models with no cosmological constant, or with $\Lambda = 0$, describe three future possibilities for the Universe with no dark energy (*thin solid lines*). It can become closed with positive space curvature, forever open with negative space curvature, and always open and no curvature in space described by Euclidean geometry. In recent times, this last option, of never ending expansion in space without curvature, has been modified by a non-zero cosmological constant to give a boost to some age estimates and permit an accelerated expansion by a mysterious dark energy (*dashed lines*).

Still, the insertion of an *ad hoc* adjustable parameter makes Einstein's theory more complex and removes some of its beauty and elegance. And the cosmological constant seems so arbitrary, being introduced every time observations contradict theoretical expectations. In addition, it is now supposed to represent dark energy inherent in empty space, but the *General Theory of Relativity* doesn't show what this energy is, where it came from, or why it should be filling space, devoid of any matter and radiation.

The Quintessential

Around 350 BC, the Greek philosopher Aristotle rejected the notion of empty space, arguing that an object moving in such a void would have no sense of direction, and wouldn't know where to go. And since it would meet no resistance, an object might move impossibly fast in empty space, even at infinite speed.

Aristotle also proposed that our imperfect, decaying material world is made of four elements – earth, air, fire and water, while the perfect, everlasting stellar heavens are made of an unchanging fifth essence, or quintessence. And that's the name given by modern scientists to a sort of invisible haze or fog that might suffuse space, filling all its nooks and crannies while also causing the expansion of the Universe to accelerate.

To produce the observed dark-energy effect, the quintessence mimics the energy field that supposedly caused the inflation, or accelerated expansion, of the early Universe. Except that quintessence is much weaker than inflation, so it takes much longer to produce noticeable effects.

At least for the time being, there is a serious disconnect from the observable world, with little or no testable consequences for the hypothetical quintessence. The theoretical physicists don't know where quintessence comes from, except that it could arise from extra dimensions, which might not exist and if they did could not be seen. At least so far, quintessence is just another name for the unknown.

So now we have two tooth fairies in the form of dark matter and dark energy. [A tooth fairy leaves money to sleeping children in exchange for a tooth that has come out, but she is about as real as Santa Claus.] First dark matter controlled the fate of the Universe, now its mainly dark energy (Fig. 5.14). Dark matter might have been the dominant factor back at the beginning, when galaxies first formed, and it might become important in the distant future if dark energy diminishes in strength.

The difficulty is that we know almost nothing about either dark matter or dark energy. Neither one of them has ever been detected directly, and scientists don't have the slightest idea what either one of them are. They have just quantified our ignorance, telling us how much we don't know.

The verdict therefore isn't in on either dark matter or dark energy and the situation remains unsettled. Other unknown forces might be involved, or perhaps our understanding of gravity is incomplete. If so, the entire "dark" edifice could collapse like a house of cards, and future scientists may find that all this dark stuff was imaginary conjecture, pure speculation, remembering that their mothers told them not to go out and play in the dark at night.

Today's cosmologists are nevertheless ever resilient and imaginative, adept at reshuffling their theories and re-evaluating reality when confronted with new, conflicting observations. Some of their models now incorporate inflation, the cosmological

FIG. 5.14 Dark mass and dark energy constraints Three independent sets of observations provide constraints to the mass and energy content of the Universe – high-redshift supernovae, galaxy cluster inventories, and the cosmic microwave background radiation, abbreviated CMB. Studies of high-redshift Type Ia supernovae constrain the difference between the densities of matter and the density of dark energy in the Universe. Anisotropies in the cosmic microwave background radiation constrain their sum. By observing galaxy clusters at widely ranging distance with the *Chandra X-ray Observatory,* the relative densities of dark matter and dark energy can be bracketed. Here the densities are given in terms of the omega parameter, denoted by Ω. It is the ratio of the inferred density to the critical mass density needed to stop the expansion in the future. The subscript Λ denotes the cosmological constant, a possible form of dark energy, while the subscript m denotes matter. Galaxy observations indicate a mass density of at most $\Omega_m = 0.3$. The theoretical expectation of an inflationary Universe without spatial curvature requires $\Omega_m + \Omega_\Lambda = 1.0$, and significant dark energy with $\Omega_\Lambda = 0.7$. The supernova, CMB and galaxy cluster observations are consistent with such a Universe, described by ordinary Euclidean geometry. (Courtesy of the Supernova Cosmology Project and Saul Perlmutter.)

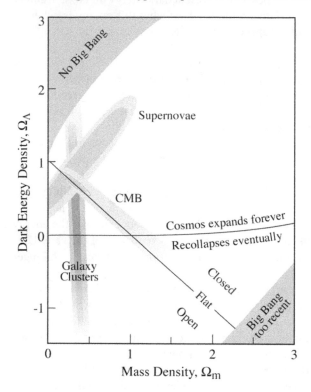

constant, and invisible particles known as cold dark matter, but these components lack comprehensive observational verification, either individually or collectively. So now it's a good time to step back and appraise all those fascinating things that we know with much greater certainty.

5.8 WHAT WE HAVE LEARNED SO FAR

Astronomers have begun to part the Cosmic Veil in an ongoing search for the unseen, providing us with a partial glimpse of the Universe in its entirety. It is an unfolding drama made possible with new technology and novel telescopes, which have often been a byproduct of military considerations. These instruments have revealed a hidden, invisible Universe with a richness and complexity that no one ever imagined.

Knowledge of nuclear reactions has shown us how stars shine, and fulfilled the alchemists' dreams by demonstrating that the elements were forged in the Big Bang and within stars. Cold War rivalry between the United States and the Soviet Union resulted in astronauts planting an American flag on the Moon, inaugurating the Space Age in which the planets and their satellites became the new frontier, the ultimate in unexplored territory. The inquisitive eyes of robotic spacecraft have transformed them into fascinating, diverse worlds. And a multitude of giant planets have recently been discovered in close orbit around nearby, Sun-like stars, laying the foundation for the future quest of smaller, Earth-sized planets that could harbor life.

We've noticed that everything moves, with nothing at rest, and gradually extended the Cosmos from the stars to billions of remote galaxies, all rushing away from us and from each other. This expanding Universe of galaxies is completely invisible to the unaided eye, requiring powerful telescopes to collect their faint visible light. The galaxies are spinning, darting and flowing through the Universe, pulled here and there by massive concentrations that distort the uniform expansion. Wherever we look, the galaxies are also lumped together into vast chains and walls, marking the edges of gigantic, empty voids, even billions of light-years away.

Radio and X-ray telescopes have revealed an explosive Universe that lies beyond the range of visual perception. It includes exploding stars, neutron stars, pulsars, black holes, radio galaxies, quasars, and the extraordinarily powerful gamma-ray bursts. Radio astronomers have even discovered the relic radiation of the Big Bang, and used its rippling temperature fluctuations to measure the geometry of the Universe and inventory the dominant types of mass that helped the Cosmos grow.

We've also learned that the emptiness of space is an illusion. The transparent, outer atmosphere of the Sun is continually blowing past the Earth and filling interplanetary space, and the Sun's gravity curves the space around it. The black spaces between the stars are filled with unseen atoms and molecules detected by their radio signals. The gravitational force of dark, invisible material is required to hold individual galaxies and clusters of galaxies together. Scientists have even speculated that all the mass and energy in the Universe could have originated from the nothingness of empty space, during a rapid inflation soon after the Big Bang or by a host of tiny, vibrating strings at earlier times closer to the Big Bang. And an enigmatic dark energy has been discovered that counteracts the mutual gravitational attraction of all the matter in the Universe, accelerating its expansion. The mysterious dark energy is supposed to permeate all of space, and to arise out of its nothingness, but its detailed properties remain unknown.

These remarkable discoveries have provided us with a vastly extended sense of the Universe and our place in it. New worlds, invisibility, motion, content, form, impermanence, violence, and emptiness – these are vehicles for interpreting both the human condition and the Cosmos, helping us to take the Universe personally. They've strengthened astronomers' resolve to further part the Cosmic Veil, confident that the Universe remains pregnant with unexpected possibility. And they have touched on questions of origins and destinies, the beginnings and ends of both the Universe and humans within it.

Chapter Six
Epilogue – Origins and Destinies

6.1 NOTHING LASTS, EVERYTHING CHANGES

Impermanence

It was once thought that the observable Universe was unchanging and eternal, but now we know that it is impermanent. There is nothing in the Cosmos that escapes the ravages of time. We notice the seemingly endless transformation in the changing seasons, the passage from youth to old age, the erosion and renewal of mountains, the life and death of stars, and the history of the expanding Universe.

Change might even be the essence of life, the only thing you can count on. Writers have dwelled on it, such as Marcel Proust's *In Search of Lost Time* and Kazuo Ishiguro's *The Remains of the Day*.

The passage of time is also portrayed in paintings, such as Georgia O'Keefe's studies of growth and decay – her luxuriant flowers, parched bones, and canyons with their geological layers of wind- and water-shaped rocks. And there is a captivating work by Paul Gauguin at the Museum of Fine Arts in Boston, which he entitled "D'où venons-nous? Qui sommes-nous? Où alons-nous?" or "Where do we come from? Who are we? Where are we going?" Beautiful, nut-brown people are carefully placed, apparently pondering the meaning of life with a pensive air, and the epic cycle of birth, life and death unfolds across the painting with a sleeping girl, a beautiful young woman and an old woman.

In just a few decades, everything that now seems so important to us will have ceased to exist and become part of something else. The Buddha reminded people about that endless change 2,500 years ago, declaring:

All that has been gathered will disperse,
All that is constructed will decline and fall,
All that meets will one day separate,
And all that lives will vanish into death.[74]

Everything is engaged in an ongoing metamorphosis, changing from one form to another, arising as one thing and passing away as another thing. This transformation, this continued rebirth, can have great significance. In Thomas Carlyle's words, "all destruction, by violent revolution or however it be, is but new creation on a wider scale."[75]

Dynamic Earth

Even the great continents, mountains and oceans are changing with time. Their permanent appearance is an illusion caused by the brevity of the human life span. The continents disperse and then reassemble, over and over again, roaming and wandering about the planet in an endless journey with no final destination. And it is only a matter of time before a colossal mountain will be eroded away into sediment and dust, or built anew when continents collide and weld together. The moving continents can even squeeze an ocean out of existence.

New ocean floor spills out of volcanoes at the bottom of the ocean, spreading sideways and helping propel the moving continents. The spreading sea floor eventually dives back into the Earth, dissolving into molten rocks and producing volcanoes that make continents grow at their edges.

The Sun Will Fade Away

In a few billion years, the Earth will become uninhabitable, when the gradually brightening Sun becomes hot enough to boil the Earth's oceans away, leaving the planet a burned-out cinder, a dead and sterile place. Eventually, all life in the Solar System is doomed. After a relatively brief existence as a hot, extended giant star, the Sun will collapse into itself, squeezing its enormous mass into an insignificant cinder about the size of the present-day Earth. Nuclear reactions within the Sun will then be a thing of the past, and there will be nothing left to warm the moons or planets. The former Sun will just gradually cool down and fade away into old age, plunging all of the planets into a deep freeze.

Dying Stars

The stars seem immutable, but they are not. They are all impermanent beacons, destined to vanish into the darkness. Some have relatively long lives, others will expire sooner, and none of them is eternal. Dim stars of Sun-like mass have settled down to a long, rather uneventful life lasting billions of years. Some of the brighter, massive stars, the red giants, have advanced to the next stage of stellar life, and the exceptionally luminous stars, with the greatest mass, have already run out of fuel and descended back into the dark, while littering space with heavy elements that were synthesized inside their cores. These chemical elements are the building blocks of future planets and the bodies of living things.

The British biologist and author Julian Huxley has captured this vision of dying stars with:

> And all about the cosmic sky,
> The black that lies beyond our blue,
> Dead stars innumerable lie,
> And stars of red and angry hue
> Not dead but doomed to die.[76]

Or the American John Updike, writing in his short story *The Astronomer*: "what is the past, after all, but a vast sheet of darkness in which a few moments, pricked apparently at random, shine?"

The Universe is Changing

By peering out at galaxies that are located at vastly different distances, astronomers have shown that the Cosmos has a history, and that the properties of galaxies change over cosmic time scales. The light we detect from a remote galaxy has traveled for a very long time, and was emitted in its infancy. And when our telescopes detect the light from a nearby galaxy, it was generated a relatively short time ago, after the galaxy has aged for billions of years.

Like children, the galaxies are more active in their youth. Distant, young galaxies shine with the intense blue light of new stars. They are forming stars at a much greater rate than nearby, older and redder galaxies. And it is expected that the truly ancient galaxies will expend all their resources and eventually fade from view, as in Robinson Jeffers' poem:

> I seem to have stood a long time and
> watched the stars pass.
> They also shall perish I believe.
> Here today, gone tomorrow, desperate wee
> galaxies.
> Scattering themselves and shining their
> substance away
> Like a passionate thought.[77]

The super-massive black holes at the centers of young galaxies consume the surrounding gas and stars, feeding the powerful jets of radio galaxies and quasars. But due to the dwindling supply of in-falling material, the central black holes of older galaxies are inactive. That explains why quasars, located in the cores of galaxies, are more common at great distances than nearby.

In both the Universe and humans, youth means greater activity, and the young are also hotter. The hot, luminous quasar spurt of activity happened a long time ago, in the distant past, shortly after the first galaxies were born, and it does not occur today. Further back, at the beginning of time, the entire observable Universe was compressed to high density and heated to enormous temperatures; now it is relatively cool and empty.

Where Did the First Stars Come From?

About 14 billion years ago, when the Big Bang set today's Universe in motion, the Cosmos was without form. There were no stars or galaxies. Somehow the smooth, unruffled fireball of the Big Bang became lumpier and clumpier, congealing into the first stars, and the observable Universe emerged. But just how this happened is a matter of ongoing controversy. Cosmologists have often thought the solution was in their grasp, but their theories change at a rapid pace.

Back in the 1970s, cosmic strings of energy, created during the birth of the Universe, were imagined to snake though space, providing the attraction needed to pull the stars and galaxies together. But the speculation turned out to be no more relevant to cosmic origins than a spider web. And other theories, like supersymmetry, super-gravity and superstrings, rose to take the place of the cosmic strings, and they were all destined to fail. In other words, most theories for the origin of the Cosmos just turn out to be wrong.

Modern cosmology indeed sometimes takes on the aspects of a new mythology, a new religion. Some cosmologists even claim that they have discovered how the Universe began. Yet, science provides only limited and partial answers to the questions of origin and destiny.

6.2 ORIGIN OF THE UNIVERSE

The Mystery of Creation

Myths are stories and metaphors that convey the wonder of the world, in the human search for truth, meaning and significance. And every culture has its myth of the creation of the world, rooted in human experience. It might be the hatching of a cosmic egg, or the growth of a cosmic seed. But where the first egg came from, and who or what planted the first seed, remained a mystery.

All the great religions also have their account of creation. They invoke a God or gods to describe the beginnings, which lie beyond the visible world.

Both the Buddhist and Hindu religions envisage an endless cosmic cycle in which the Universe has always existed and will always exist, consisting of cyclical and eternal worlds that are forever undergoing creation, destruction and rebirth, arising and disappearing over and over again forever. The Hindu god Shiva, in the form of the Lord of the Dance, represents this rhythm in a Cosmic Dance surrounded by a circle of flames, a ring of fire (Fig. 6.1). One of Shiva's hands holds a drum, whose sound is the sound of creation; another hand holds the fire of destruction, a reminder that the newly created world will eventually be obliterated.

In the Judeo-Christian view, found in the *Bible* at the first *Book of Genesis*, the Universe had a definite beginning, when God created heaven and Earth. The English engraver and poet William Blake drew the shaping of the round Earth by the sweeping compasses of God (Fig. 6.2), while his poet colleague John Milton wrote, in *Paradise Lost, Book VII, lines 224 to 227*.

> Then stayed the fervid wheels, and in his hand
> He took the golden Compasses, prepared
> In Gods eternal store, to circumscribe
> This Universe, and all created things.

The all-powerful, supernatural God was the source of the entire observable Universe, itself non-recurrent and unidirectional in time.

Now, a vocal minority of cosmologists has pushed into the domains of myth and religion, at the boundary between what is known and the mystery of creation. Mathematical equations and computer simulations are employed to construct their own epics, using the language of science rather than mythology. They have attempted to show how everything – mass, energy, space and time – might have originated in the Big Bang that set the Universe in flight, and then moved on to speculate about what happened before the beginning. They have become today's shamans, hiding their

FIG. 6.1 **Cosmic dance** The Hindu god Shiva is shown executing a cosmic dance, symbolizing a rhythm that assures and perpetuates the cyclic creation and destruction of the world. Shiva dances in the middle of a circle of fire that represents the eternal flux of the Universe, where nothing is permanent. The flame in his upper left hand (*right side*) symbolizes destruction; while creation is energized by the primordial beat of the drum held in his upper right hand. His other right hand (*lower left side*) has the palm held out in a gesture of reassurance and protection. Shiva's right foot (*not shown here*) tramples a dwarf who symbolizes ignorance and evil. This sculpture is located in a magnificent temple built in the late 7th century at Kanchipuram, in southern India, whose name is derived from the words *Ka,* for creator, and *anchi,* for worship. (Photograph by Kenneth R. Lang, 2001.)

magical pronouncements behind equations or even books that almost no one can read or understand. With an amazing arrogance, presumption, hubris and lack of humility, they even claim authority because of it, asserting that they alone can perceive the true beginnings of the Universe. But that does not make their explanations true.

Before the Big Bang

How did the observable Universe come into being? Where did everything come from, and why is there something rather than nothing at all? In attempts to answer these great mysteries, astronomers have peered out into space and back into time, to show that the first stars and galaxies originated far away and long ago.

They have run the expansion of the Universe backward into a hotter and denser past, tracing the history of the Universe to the first moments of the Big Bang, and then followed it back again into the present, to show how the chemical elements were synthesized. This has been one of the most productive and satisfying exercises in the history of cosmology, convincing even the skeptical that the observable Universe started with a Big Bang that occurred about 14 billion years ago.

FIG. 6.2   The ancient of days The English painter, engraver and poet William Blake (1757–1827) used this image as the frontispiece of the illustrated book *Europe: A Prophecy*. The figure is very probably Urizen in the act of creating and/or circumscribing the material Universe. Various persons have titled the image as "The Ancient of Days", "God Creating the Universe" and "God as an Architect". God is represented as an old, but muscular, man with long hair and beard, squatting in a circle. He holds a pair of dividers or compasses in his left hand, indicating the act of creation and design. Beams of light emanate from the circle, suggesting that it represents the Sun. (Courtesy of The Whitworth Art Gallery, The University of Manchester, Great Britain.)

But when astronomers push back to before the Big Bang, they run into a barrier to knowledge. The ultimate frontier, the very beginning, always remains shrouded in mystery. No one knows what created the Big Bang, why it occurred, or whether or not anything was happening before it took place.

The reason for this breakdown in knowledge is a failure of the known laws of science at the very beginning of the observable Universe, at the peak of the cosmic inferno. When extrapolated to the limit, to that crucial first instant, the equations of Einstein's *General Theory of Relativity* blow up into infinities of density, temperature, and the curvature of space-time. Einstein himself realized the dilemma, reasoning that his equations might not be valid at very high densities.

Since everything was compressed to an exceedingly small size near the beginning, scientists naturally tried to patch *General Relativity* up with *Quantum Mechanics,* which was not described in the original theory. This added an element of uncertainty, the notion of a fuzzy, probabilistic fog, to the first moments, and that was something that Einstein distrusted. "Quantum mechanics is very worthy of regard," he wrote Max Born, "but an inner voice tells me that that this is not yet the right track. The theory yields much, but it hardly brings us closer to the Old One's secrets. I, in any case, am convinced that *He* [God] does not play dice."[78]

Cosmologists have been trying to forge a link between two bastions of modern physics. *Quantum Mechanics* had been well tested at the sub-atomic level. *General Relativity* predicted the Sun's observed curvature of nearby space, and was consistent with the observed expansion of the Universe, interpreted as a stretching of space-time.

Yet, in spite of more than half a century of trying, the two theories have yet to be linked together into a conclusive, tested explanation for the beginning of the observable Universe. So we still do not know our cosmic origins, let alone what might have existed before the Big Bang.

Some scientists say we will never have direct evidence for what happened. If there were an accelerated expansion in the first miniscule moments of the Big Bang, this inflation might obliterate any evidence of previous space, time and matter, erasing previous history. That cosmic forgetfulness closes the door to the beginning, conveniently removing the event from any observational consequences. And this lack of hard evidence has resulted in many imaginative speculations.

An eternally existing, self-reproducing inflationary Universe has even been imagined, in which inflationary bubbles endlessly clone other ones. Our Big Bang might not have been the only one, and our Universe could be just one small part of a multitude of possible Universes that might have bubbled up out of vacuous nothing. But we can't see any of these other bubbles, just the one we are immersed within. And perhaps the only Universe that we can see is one compatible with our existence.

All of these other hypothetical Universes are so widely separated that they are beyond communication with, and completely disconnected from, each other. They are forever invisible, occupying places we cannot see, the ultimate in un-testable speculations.

It's a fascinating, imaginative idea, but then there might also be fairies and elves prowling about your house at night. And it's not an entirely new concept. Over 250 years ago, Thomas Wright supposed that a plentitude of Universes had been spawned, each centered on its own Divine Presence (Fig. 6.3). Recent attempts involve more mathematics, but it is

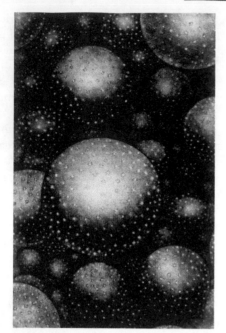
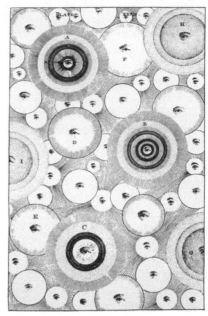

FIG. 6.3   Multiple Universes Thomas Wright of Durham (1711–1786) supposed that the Milky Way is just a small segment of a spherical shell of stars, with the Sun located in the shell and the shell centered upon a Divine eye (Fig. 3.6). Wright also supposed that there might be many such Universes, each centered on its own Divine Presence, illustrating his ideas with these drawings. (Adapted from Thomas Wright's *An Original Theory or New Hypothesis of the Universe,* published in 1750, reproduced by Science History Publications, New York 1971, pages 173 and 176 plates XXXI and XXXII.)

still just as likely, from the scientific point of view, that there is only one Universe, the one in which we find ourselves and can see.

If we probe further back into time, before the hypothetical inflation that created our personal Universe, the energies become higher and higher, perhaps entering the domain of the tiny, invisible vibrating strings that occupy unseen places and extra dimensions, springing out of nothing and accounting for everything. The putative strings are supposed to unify *General Relativity* and *Quantum Mechanics,* and some say they can show how the Big Bang was an outcome of a preexisting state. Yet, despite a quarter century of investigation, the string experts still do now know how to test their theory, or even how to solve the equations in the extreme conditions at time zero.

Many astronomers, who like to observe the real world, are becoming fed up with all the hype, the persistent appeal to the unobservable and unverifiable, created from complex mathematics and esoteric theory that almost no one can understand. As the American writer Mark Twain quipped: "There is something fascinating about science. One gets such wholesale returns of conjecture out of such a trifling investment of fact."[79]

Cosmology, Science and Religion

Cosmologists deal with the exotic topics, about the beginning and end of the Universe as a whole. So their pronouncements appeal to the public at large, who are naturally concerned about the related events in their lives. Now, armed with an arsenal of new observations, from the large-scale structure of galaxies to ripples in the cosmic microwave background radiation, the cosmologists trumpet that they have entered a new age, where their speculations are anchored in reality, constrained and made credible by observational tests.

Many observational cosmologists are indeed gathering data to test theoretical conjectures at the interface between the cosmic background radiation and the first stars and galaxies. And cosmological nucleosynthesis at say three minutes after the Big Bang has been well tested.

Nevertheless, hypothetical theories of a few modern, highly educated cosmologists are still often disconnected from the observable world, with little or no testable consequences and without objective ways to verify their claims. And should some observation cast doubt on someone's favorite theory, it is explained away by adding an extra parameter to the equations, with endless adjustments to "save the phenomenon."

The Earth-centered cosmology of ancient times was such a paper world. Ptolemy and others used all sorts of mathematical acrobatics to retain the beauty of simple circular motion. And then Kepler derided their attempts, replacing the elegant circle with a coarser ellipse and a Sun-centered cosmology. Observations, from Galileo on, showed that Kepler was right. Even Plato, a lover of circles, opinioned "I have hardly ever known a mathematician who was capable of reasoning."[80]

To be polite, modern cosmology is at best premature. And to be honest, lets say that an outspoken minority of cosmologists are operating a cosmic con game, worthy of J. T. Barnum and his Greatest Show on Earth. They are deluding the non-scientists, who are ill equipped to judge the veracity of their claims, and cannot even discriminate between the "legitimately weird and the outright crackpot."[81] As Barnum was fond of saying, there is a sucker born every minute, and in this case the suckers are the popular press, the uneducated public and governmental funding agencies.

For decades now, high-energy physicists and cosmologists have claimed to be on the threshold of the ultimate theory, a complete explanation of everything, even invoking terms like the God particle. The British black hole expert Stephen Hawking, for example, concludes one of his books with: "Up to now, most scientists have been too occupied with the development of new theories that describe *what* the Universe is to ask the question *why*… However, if we do discover a complete theory, it should in time be understandable in broad principle by everyone, not just a few scientists. Then we shall all, philosophers, scientists, and just ordinary people, be able to take part in the discussion of the question of why it is that we and the Universe exist. If we find the answer to that, it would be the ultimate triumph of human reason – for then we would know the mind of God."[82]

Well, the dream of a final theory has not been realized. No scientist knows how our Universe came into being, and its origin may remain forever outside the scope of science, most likely overlapping religion. Some mysteries, such as the origin of the Universe and life, may not be penetrated by science alone.

For most of history, cosmology has indeed been part of mankind's religious rather than scientific view. For thousands of years Buddhism has taught that the Cosmos we live in is only one of an infinite number, and that the Universe never begins or ends, but endures. About 1,600 years ago, Saint Augustine, the sinner turned Christian philosopher, proclaimed that there is no "before" or "after" in God, as there is in creatures, and that time began with his creation of the world.[83] Howard J. Van Till offers a succinct summary of St. Augustine's literal interpretation of the book of *Genesis* in the *Bible* with these words:

> The Universe was brought into being in a less than fully formed state but gifted with the capacities to transform itself, in conformity with God's will, from unformed matter into a truly marvelous array of physical structures and life forms.[84]

In the 1930s, the Catholic priest Georges Lemaître applied Augustine's ideas to the origin of the expanding Universe, proposing that the Universe was made *with* time and not *in* time, so there was no "before" the Big Bang. And since the creation event coincided with the appearance of space, it did not occur in space, at some particular location. So there might be no meaning to the questions of what happened before the Big Bang, or what the Universe is expanding into.

As Lemaître realized, radiation was the most powerful thing around in the first moments of the Big Bang, and this idea wasn't that different from the book of *Genesis*, in which God said, "Let there be light." So the Catholic Church found no disagreement with Lemaître, and in 1951 Pope Pius XII endorsed the new cosmology, arguing that it demonstrated that the Universe had a beginning in time and proved the existence of God, the Creator.[85]

Science is itself a spiritual undertaking, initiated with theories that are non-proven beliefs and operating under a faith that nature can be explained by objective laws, which are verified by tests in the observable world. Both science and religion are quests for a deeper understanding of the mysterious unknown, attempts to tap into something that normally operates beyond the range of known perception. Neither has a monopoly of truth.

The great physicist Albert Einstein proposed a union of science and religion, believing that the religion of the future will be a cosmic religion, encompassing both the natural and the spiritual.[86]

Religions will always address questions that may not be answered by direct scientific inquiry. What is the purpose of the Universe? What is the meaning of human life? What happens after death? How did life originate? Science is most likely incapable of answering these basic questions about human existence.

6.3 THE BEGINNINGS OF LIFE

Conditions for Life's Beginning and Survival

This marvelous, pulsating thing we call life. Where did it come from, and where can it be found? And what is life anyway? Answers to these questions are at the very limits of scientists' capabilities.

Their approach to life is biological and chemical. The atomic and molecular constituents of living things are identified, and the conditions required for their production and survival are delineated.

Every living thing on Earth is composed of organic molecules, which are based on carbon atoms that form large molecules by combining with other atoms, such as hydrogen. These organic molecules provide terrestrial life with the capacity to adapt to a changing environment and to replicate itself.

Liquid water is also crucial to life here on Earth. We ourselves are largely composed of water. And just about everywhere there is water here on Earth, there is some sort of life.

Perhaps the most important key to life is energy, which flows through life and powers it. All living things have to extract energy from their environment to fuel themselves. And the most powerful source of this energy is sunlight striking the surface of the Earth.

Photosynthesis by plants, for example, uses the energy of sunlight to extract carbon from carbon dioxide molecules in the atmosphere, incorporating the carbon into organic molecules with the help of water.

Heat energy from the Earth's hot interior can also energize life. Subterranean and deep-sea organisms substitute volcanic fires for solar ones as an energy source, living off the Earth's inner heat and inorganic chemicals.

To stay alive, organisms also have to be protected from their own waste products, and cells accomplish this. The outside membranes of cells wall off their interiors from their own chemical byproducts.

From an astronomer's perspective, planets and stars are also necessary, but not sufficient, conditions for life. A life-sustaining planet requires a solid surface, liquid water and a nurturing atmosphere. For humans, this means a rocky surface to walk and develop on, and an atmosphere with lots of oxygen to breathe. On the other hand, plants can float in water, and breathe carbon dioxide, returning oxygen into the air.

A star is also required to heat and light a planet, which has to be at just the right distance for liquid water to exist. Although oceans of liquid water cover most of the Earth, any water on the planet Venus, which is just a little closer to the Sun, was boiled away long ago, and almost all the water on the surface of Mars, a little further away from the Sun, is now frozen into ice.

So sunlight warms the world, keeping most of Earth's water liquid, and liquid water is vital for life. Furthermore sunlight energizes most life on Earth. Stars that are dimmer than the Sun, and the large majority of stars are, can only warm a planet in very close orbit, where the star's gravity would force one planetary hemisphere to face the star and the other to turn away, boiling any water on one side and freezing it on the other. And a star that is significantly brighter than the Sun might not live long enough for life to develop on one of its planets.

Origin of Life on Earth

Our understanding of even the possibility of life on other worlds is hampered by a lack of knowledge of how life originated on Earth. Fossil records suggest that some sort of life existed between 4.0 and 3.5 billion years ago, but earthquakes and volcanoes, moving and colliding continents, erosion by running water, and even the oxygen produced by life itself, have erased the clues to the origin of this early life on Earth.

One speculation, initiated by the Russian biochemist Alexander Oparin in the 1920s, proposes that life emerged in the shallow seas or tidal pools of the young Earth, which might have been enveloped by a warm, moist, hydrogen-rich atmosphere. Given enough time, simple organic molecules found in the primitive atmosphere would participate in chemical reactions in the liquid water, energized by ultraviolet sunlight or lightening. Ultimately complex organic molecules capable of reproducing themselves might have formed, even within tiny cells bounded by membranes.

In the 1950s Stanley Miller showed that an electrical discharge in a flask-bound rendition of a primitive, hydrogen-rich atmosphere of methane, water and ammonia can give rise to amino acids. Since that time, a number of experiments have been performed in which the simulated atmosphere is converted by electrical discharge or energetic radiation into more complex organic molecules found in living things.

However, the experts are not so sure about the composition of the atmosphere that enveloped the young Earth. It might have contained substantial amounts of water vapor, carbon dioxide, the main ingredient of the current atmospheres of Mars and Venus, or nitrogen, the dominant molecule now in the Earth's atmosphere. The production of organic molecules like the amino acids is more difficult in this case.

An alternative speculation proposes that the first pre-biotic organic molecules originated deep beneath the surface of the Earth. This theory is related to the discovery in the late 20th century of microbial organisms that now thrive on Earth under extreme conditions that were once thought to be lethal. Some of these extremophiles, as they are called, exist at temperatures near and above the boiling point of water; they would freeze to death at temperatures that would result in severe burns to humans. Other microbes now exist at below the freezing temperature of water, underneath and within Antarctica sea ice.

Bacteria have even been found that thrive on the gas given off by raw sewage, living in concentrations of acid strong enough to dissolve metal. Other tiny organisms proliferate under dark, pressure-cooker conditions at the bottom of the ocean, feeding on materials emerging from volcano vents, and some of them dwell within the Earth's rocky interior more than a kilometer down.

Instead of looking within the continents or oceans, we might have to look out to the sky for the source of life. About 40 thousand tons, or 40 thousand kilograms, of cosmic dust falls from space onto Earth in a typical year. Some of these dust-like particles originate from comets, others from asteroids and meteorites. The tiny dust particles don't burn up as they fall through the air, but float down to the ground.

This cosmic dust is everywhere. It is in the air we breathe, the food we eat, and the water we drink. Some of the dust that has been spawned by comets and asteroids even enters your hair. A few astronomers have endowed the dust with life-giving properties, bringing self-replicating carbon molecules to Earth.

Carbonaceous meteorites abound with organic compounds, including at least 20 types of amino acids, and this implies that these molecules probably existed in the Solar System since it originated, perhaps before terrestrial life began. But the organic molecules found in meteorites are non-biological, and they are not in themselves vestiges of extraterrestrial life. Such molecules *might* nevertheless have been the chemical precursors to living matter.

So the building blocks of life might have come from space, and some scientists propose that life itself could have traveled here. Almost a century ago, the Swede Svante Arrhenius, a Nobel Prize laureate in chemistry, proposed that tiny microorganisms, the spores, could travel across space, from other planets or stars. Single-celled organisms, for example, could have developed on Mars before being thrown into space to come and fertilize the Earth.

It is possible that the Earth and Mars have regularly exchanged microbes over the years. Cosmic impacts with Mars have sent hundreds of tons, or hundreds of thousands of kilograms, of nomadic Martian rocks to Earth over the centuries. And comparable amounts of terrestrial rocks have smashed into Mars. So, life on Earth could have come from Mars, and in that case we might all be Martians. Or life might have originated on Earth and was then delivered to Mars. Maybe life arose on the two planets independently and spontaneously, or perhaps impacting comets or asteroids pollinated both planets. And, just maybe, Earth is the only place to harbor life in the entire Solar System.

So many theories, so little direct evidence; it's a sure sign of uncertainty. Scientists don't know for sure just where the molecular ingredients of life came from, let alone how living things originated from them. The chemicals are just supposed to have spontaneously come alive, and life began.

The beginning of life on Earth thus remains a captivating mystery. And perhaps because scientists cannot explain life's origin on Earth, they are seeking additional clues by looking for signs of life in other places.

Searching for Life Elsewhere

Mars is the most likely planet to harbor extraterrestrial life, either as living organisms or as fossils of former ones. The red planet is now much colder and drier than Earth, and it does not have enough heat, liquid water or atmosphere for most, or possibly all, forms of terrestrial life. Nevertheless, microbes can survive under very extreme conditions on Earth, so they might also exist on Mars.

The surface temperature on Mars is usually below the freezing point of water, and most of its water is therefore frozen solid. And there is a lot of it; orbiting spacecraft have detected vast amounts of water ice just below the Martian surface. Internal heat might warm the ice to liquid water inside the planet, but if it reached the surface, the water would either evaporate or freeze. So it cannot now rain on Mars, and liquid water can only briefly survive on the surface. In addition, Mars' atmosphere is now a hundred times thinner than our air, consisting mainly of carbon dioxide and containing no ozone layer to protect the surface from the Sun's deadly ultraviolet radiation.

Despite the harsh present conditions, life might have thrived during more clement conditions on Mars eons ago. A vast deluge of liquid water once swept across the planet's surface, carving out immense flood channels and ancient river beds that are seen today. And roving spacecraft have obtained abundant evidence of flowing water in the past; including minerals that require prolonged seepage in liquid water to form.

So Mars was probably warmer, wetter and more hospitable to life in the distant past, with a thick, dense atmosphere. Robotic vehicles or astronauts might find the fossilized remnants of life that thrived back then. Living organisms could even have survived until now, but the prospects are dim, at least on the planet's hostile surface.

In 1976 the *Viking 1* and *2* spacecraft landed on the surface of Mars, and tested the soil for signs of living or dead microorganisms. They found no detectable organic molecules, which meant that there are no cells at the landing sites, living, dormant or dead. If they ever existed, the lethal soil wiped them out, while also rusting Mars red.

Ancient Martian life might have survived not on, but beneath the surface, migrating below the chemically reactive soil. Overlying rocks would provide shielding from the deadly ultraviolet radiation. And the internal heat might provide the energy needed to melt the ice, and produce chemical reactions that could sustain tiny microbes.

Thus, all the basic ingredients for life may have once been present on Mars, and life *could* have arisen there. And living microbes *might* still be found inside the rocks or underground. They would be very small, even microscopic, certainly not the creeping, crawling variety. But these are hopeful speculations, and we won't know for sure until robotic spacecraft and astronauts visit Mars and find out.

The discovery of life on Mars, even primitive life in the very distant past, would have profound implications. It would give us companionship in a vast and lonely Universe, and it would also be a little humbling. The discovery would raise the likelihood that life might be found elsewhere in the Universe as well, either on the moons of Jupiter or Saturn, or on one of the countless planets that surely exist in our Galaxy (Focus 6.1).

Someday astronomers might find that the English poet Alexander Pope was prophetic, writing in his famous *An Essay on Man*:

> He, who through vast immensity can pierce,
> See worlds on worlds compose one Universe,
> Observe how system into system runs,
> What other planets circle other suns,
> What varied Being peoples ev'ry star,
> May tell why Heav'n has made us as we are.[87]

There are roughly 100 billion (10^{11}) stars in our Milky Way Galaxy, and the number of planets could outnumber the stars. So there ought to be at least one habitable, Earth-like planet out there. It could be swarming with organisms, teeming with life. Given some water, energy and simple molecules, which are found in copious amounts in interstellar space, chemical reactions ought to create the relevant molecules and perhaps even microbes. It's statistical, it's probable, and it's likely. So you might not be alone in the Cosmos. But you won't be certain until someone finds life elsewhere. Even on Earth, with its 6.5 billion people, there similarly ought to be at least one person ready to love you, but you won't know until it happens.

Believing in the existence of intelligent life in our Galaxy is quite another matter. It's very unlikely that someone else is out on another planet, lying awake at night wondering if we exist. And life, be it intelligent or not, might be unique to Earth. That would give us extra cosmic significance and responsibility.

But life could also be ubiquitous throughout our Galaxy. No one knows if there is life elsewhere, and the possibility is as speculative as the origin of either life or the Universe.

FOCUS 6.1

Widening the Search for Life

Our conjectures about life on other worlds may turn out to be too parochial or Earthlike, and totally alien life forms might exist on distant planets orbiting the Sun or other stars.

The possibility of a warm, global ocean just below the gleaming, ice-covered surface of Europa, a large moon of Jupiter, has led to speculation that life could have gained a foothold there. Liquid water apparently oozes out of cracks in the thin, smooth water-ice crust, and lubricates blocks of moving ice. Though far from the Sun, and therefore bathed in diminished sunlight, alien life might thrive in the dark, internal ocean, energized by chemicals and warmed by tidal flexing due to Jupiter's varying gravitational pull along Europa's eccentric orbit.

Saturn's giant moon Titan, larger than the planet Mercury, possesses a thick, Earth-like nitrogen atmosphere with abundant methane and other organic molecules. The entire atmosphere is well below the freezing temperature of water, but warm enough for methane to exist as a gas, liquid or ice, forming clouds and rains of methane. On 14 January 2005, the *Huygens Probe,* released from the *Cassini Orbiter,* landed on the surface of Titan, detecting signs of river channels and flowing liquid, probably methane. There is therefore speculation that Titan might resemble the early Earth in a deep freeze.

And even if life within the Solar System is only found on Earth, the discovery of planets around far-away stars suggests that life might exist in the Cosmos at large. These stars and their planets were formed in interstellar clouds that contain vast amounts of water, one of the key molecules of life, as well as all kinds of organic, carbon-bearing molecules, from formaldehyde to benzene.

Numerous Jupiter-sized planets have indeed been detected revolving in tight orbits around other stars. But these are presumably huge, gaseous bodies without a surface, and any living thing would be incinerated at the searing temperatures on most of these hot worlds, so close to their star. We are now developing the technology to divulge smaller, Earth-sized planets circling stars like the Sun at comfortable distances suitable for harboring life. These worlds will eventually be scrutinized for signs of living things, such as large amounts of molecular oxygen or other atmospheric gases that are replenished by life.

Being Alive

What is the source of life? It isn't just chemicals, molecules and energy. They tell us about life's physical ingredients, how it works, but they have not described the creation of functioning, reproducing life. No one has shown how inanimate matter might have coalesced into a living thing, or how replicating biological molecules could have been created form non-biological ones. The vital spark of life, which breathed fire into the molecules and our bodies, making life arise from non-life, has not, and probably cannot, be described by science alone.

And science leaves little room for human concerns. It doesn't tell us about curiosity, joy, sorrow, or wonder, although it can be a source of them all. These qualities of life are part of being alive. They are aspects of the beauty and rapture of the human spirit and the meaning of life in the Cosmos.

6.4 FATE, DREAM AND WONDER

Everything has an end. No love lasts forever. On the personal level, our bodies, the cars we drive, and the houses we live in will eventually be gone. You will inevitably perish. And every star that you see at night will expend all its nuclear energy some day, vanishing into the darkness.

Both astronomy and religion are concerned with human fate. For astronomers, it's related to the time when the Sun dies. A few billion years from now, a brighter Sun will vaporize the oceans away and destroy life as we know it. And then its fires will be extinguished, removing all life-sustaining light and heat from the Solar System. But humans might well destroy themselves well before this cosmic doomsday.

Almost every religion includes a notion of the end of the world. Around 1300 BC, the Iranian prophet Zarathustra, called Zoroaster by the Greeks, taught that the world would end, after a cosmic battle between good and evil. The *New Testament* of the *Bible* includes an account of the apocalypse, in which the end of the world is associated with the Second Coming of Jesus. And both Muslims and Hindus believe there will be a period of moral decay as the end approaches.

There are several cycles of world birth and death in Hinduism as there is in Buddhism. For Buddhists, the world will come and disappear, come and disappear. So eventually the world we live in becomes desert and even the oceans dry up. But then again, another new world is reborn. It's endless.

Sooner or later, everything in nature must come to an end. We can think ahead to a distant time when everything we now see in the Universe will either no longer exist or be altered beyond recognition. Time may even be running out for the entire Cosmos. Its apocalypse might be on the way. But we really don't know for sure what the ultimate destiny of the Universe is, whether it is doomed to end some day or to continue forever.

Until recently, the fate of the Universe seemed simple enough to infer. It just depends on the total mass of the Universe and its rate of expansion. If there is enough mass, gravity will pull back on the outward moving galaxies and slow their retreat, ultimately reversing the expansion and dragging the Cosmos back down into oblivion. It might then be melted down and remade in the caldron of a second Big Bang.

And if there isn't enough mass, the Universe will expand forever, perhaps at a slower pace in the future, but never stopping to turn around and fall back into itself. The Big Bang would not be repeated, and the Cosmos would be destined to endlessly expand and thin out. The poet T. S. Eliot has foretold this type of cosmic destiny with:

> This is the way the world ends
> This is the way the world ends
> This is the way the world ends
> Not with a bang but a whimper."[88]

The observational evidence has come down in favor of such an ever-expanding Universe for more than half-a-century. The density of ordinary matter in the Universe, in both visible and invisible forms, is ten times too small to ever stop the expansion.

But astronomers recently discovered that the galaxies are expanding at a quickening pace, moving faster and faster as time goes on. They are being accelerated by the anti-gravity push of a mysterious dark energy. So the fate of the Universe isn't supposed to depend on its mass anymore, but on its energy. If dark energy retains its vigor, the Universe is doomed to expand forever. But if dark energy weakens as time goes on, expending its strength, gravity and mass might take over. It's unlikely, but possible that the Universe might even collapse in the future.

We know next to nothing about dark energy. The observations could be wrong, or we might not know how to extrapolate them into the future. All possibilities are open, and any destiny is possible for the Universe. The poet Robert Frost put the choices in human terms with:

> Some say the world will end in fire,
> Some say in ice.
> From what I've tasted of desire
> I would hold with those who favor fire.
> But if it had to perish twice,
> I think I know enough of hate
> To say that for destruction ice
> Is also great
> And would suffice.[89]

On cosmic time scales multiple fates await the Universe. The Sun's light will eventually vanish, in several billion years, and in 5 or 6 billion years the nearest spiral galaxy, Andromeda or M31, will collide with the Milky Way, perhaps disrupting our entire Galaxy.

Rather than coasting gently into the night, the distant galaxies are now accelerating into invisibility. If the acceleration continues unabated at the current rate, all of the galaxies will be moving apart so quickly that they will not be able to communicate with each other in about 150 billion years, disappearing over the cosmic horizon.

Eventually, in about 100 trillion years from now, all of the interstellar gas and dust from which new stars condense will be finally used up, and new stars will cease to be born in any galaxy. All the stars will wink out, the galaxies will be swallowed by black holes, and the Cosmos will go to sleep, entering perpetual night.

But the Universe isn't finished yet, and neither are we. More immediate threats await us. Most of the time we are half awake, sleepwalking through life and unaware of all that surrounds us. The tedium of daily life drags us down, wearing away the sparkle in our eyes and even the human spirit. The relentless advance of science, and the technology that accompanies it, seem to have driven some of the magic out of the world, leaving many with blurred vision, cloudy minds, and nothing to believe in. So people in our modern era suffer from a spiritual hunger, a craving for something meaningful beyond their familiar world.

Astronomy helps satisfy this hunger, providing us with a sense of something larger than us, and carrying us out into the captivating realms of the Cosmos, far beyond our daily material concerns. It has taught us how to perceive a better world and to search for the unseen, helping us to tap into the wonder, immensity and mystery of the Universe.

But astronomy and the other sciences also progress in small steps, dividing the natural world into well-defined components. And we get the gnawing feeling that it is missing something fundamental, that our perception is incomplete. It's as if we are listening to a few beautiful notes, without hearing the complete melody. So the Universe in its entirety remains unknown.

We indeed seem to live in a dream, looking out at a mirage. Nothing is exactly as it seems. Astronomy, art, poetry, music, literature and religion can all help us to penetrate the barriers that obscure our vision, to part the Cosmic Veil. When they stimulate our imaginations, space and time seem to melt away and cease to exist, and the human spirit is set free as if in flight (Fig. 6.4). We can soar in unrestrained participation in the Cosmos, becoming aware and alive.

FIG. 6.4 Icarus The French painter Henri Matisse (1869–1954) thought that happiness is derived from never being a prisoner of anything, including success or style, and he represented freedom from imprisonment in his 1947 book *Jazz* with this cut-out entitled *Icarus,* who escaped using artificial wings fixed to his shoulders with wax. The silhouetted outline of the black, mythological Icarus seems to be pushing against the downward pull of gravity, trying to break free, soar into a bright blue sky, and set the human spirit free. In the pages next to the Icarus cut-out, Matisse describes the unique vision achieved by flying in an airplane, giving us "a perception of unlimited space in which for a moment we felt ourselves to be free." Icarus' red heart symbolizes love, another theme in *Jazz.* In the handwritten text accompanying the cut-out entitled *The Heart,* Matisse declared: "He who loves also soars, runs and rejoices; he is free and nothing can restrain him." [d] Its liberating and all-consuming nature can make one see the world anew. (Courtesy of Succession Matisse, Paris.)

---

[d] Henri Matisse (1869–1954), *Jazz.* Originally published by Tériade (Efstratios Eleftheriades), in *Editions Verve,* Paris 1947. Reproduced by George Baraziller, Inc., New York 1985 with an English translation of the handwritten French text by Sophie Hawkes.

And then we might circle back to the beginning, and start our voyage into the unknown anew. As T.S. Eliot wrote, "the end of our exploring will be to arrive where we started, and know the place for the first time,"[90] or in Mario Socrate's poem:

Ebbene, forse voi credete
che l'arco senza fondo della volta spaziale
sia un vuoto vertiginoso di silenzi.
Vi posso dire allora che verso
questa terra sospettabile appena
l'universo giá dilaga di pensieri.

And in English translation by his wife Vanna:

Well, maybe you think
That the endless arch of the space vault
Is a giddy, silent hollowness.
But I can tell you that,
Overflowing with thought,
the Universe is approaching
this hardly guessable Earth.[91]

QUOTATION REFERENCES

TEXT

1. Cecilia Payne-Gaposchkin (1900–1979). *The Dyer's Hand: An Autobiography.* In Haramundanis, Katherine (ed.), *Cecilia Payne-Gaposchkin: An Autobiography and Other Recollections, Second Edition.* New York, Cambridge University Press 1996, page 237. In the last paragraph of her autobiography, she states: "It is true that we base our work on observed facts. If nothing were observed, there would be nothing to understand. But the facts are not reality: that is something that lies beneath the facts and gives them coherence. If science, as I know it, can be described in a few words, it might be called a search for the Unseen."
2. William Blake (1757–1827), *The Marriage of Heaven and Hell.*
3. Isaac Newton (1642–1727). In Brewster, *Memoirs of Newton* (1855), Vol. II, Ch. 27.
4. Albert Einstein (1879–1955), Quoted in Banesh Hoffmann: *Albert Einstein: Creator and Rebel,* New York, Viking 1972, v. Also see Alice Calaprice (editor): *The Expanded Quotable Einstein,* Princeton University Press, Princeton, New Jersey 2000, page 261.
5. Albert Einstein (1879–1955), "What I Believe," *Forum and Century* **84,** 193–194 (1930).
6. William Wordsworth (1770–1850), *Ode: Intimations of Immortality,* from *Recollections of Early Childhood.*
7. Matthew Arnold (1822–1888), *The Buried Life.*
8. Vincent Van Gogh (1853–1890), *The Complete Letters of Vincent Van Gogh.* New York Graphic Society, Greenwich, Connecticut 1958, volume 2, page 589.
9. Vincent Van Gogh (1853–1890), *The Complete Letters of Vincent Van Gogh.* New York Graphic Society, Greenwich, Connecticut 1958, volume 3, page 58.
10. Pablo Neruda (1904–1973), *Poetry,* translated by Alastair Reed.
11. Wassily Kandinsky (1866–1944), *Xxme Siecle,* 1938, quoted by his wife Nina Kandinsky, in her notes to the 1947 English translation of Kandinsky's *Concerning the Spiritual in Art,* George Wittenborn, Inc, New York, 1947, page 10.
12. Jelena Hahl-Koch, *Kandinsky,* Rizzoli, New York, 1993, page 284.
13. Joan Miró (1893–1954), Article by James Johnson Sweeney, entitled "Joan Miró: Comment and Interview," *Partisan Review,* **15,** no. 2, page 210, February 1948.
14. Albert Einstein (1879–1955), "Principals of Research," a speech delivered at Max Planck's sixtieth birthday celebration, 1918. Reprinted in *The Expanded Quotable Einstein,* Alice Calaprice (editor), Princeton, New Jersey, Princeton University Press 2000, pages 235–236.
15. Louis Pasteur (1822–1895), Lecture 1854.
16. Arthur Stanley Eddington (1882–1944), "The Internal Constitution of the Stars," *Nature* **106,** 14–20 (1920). Reproduced by Kenneth R. Lang and Owen Gingerich (editors) in *A Source Book in Astronomy and Astrophysics 1900–1975,* Harvard University Press, Cambridge, Massachusetts 1979, pages 281–290.
17. Arthur Stanley Eddington (1882–1944), *The Internal Constitution of the Stars.* Dover, New York 1951 (first edition 1926), page 301.
18. *Bhagavad-Gita, Book 11, Sections 12, 32,* English translation by Winthrop Sargeant, State University of New York Press, Albany, New York 1984, pages 464 and 484.
19. Jules Robert Oppenheimer (1904–1967), "Physics in the Contemporary World: Enhancement of Science, with Knowledge Imparted for Man's Benefit, Second Arthur D. Little Lecture," *Technology Review,*

February 1948, page 203, reproduced in *Time Magazine*, 23 February 1948, page 94.

20. Edward Teller (1908–2003). Quoted in Richard Polenberg (editor), *In the Matter of J. Robert Oppenheimer*, Cornell University Press, Ithaca, New York, 2002, page 253.

21. John Keats (1795–1821), *On First Looking into Chapman's Homer*.

22. President John Fitzgerald Kennedy (1917–1963), Address to Congress, 25 May 1961.

23. President George W. Bush (1946–), Address at NASA Headquarters, Washington, D.C., 14 January 2004.

24. Metrodorus of Chios (lived fourth century BC). As quoted from Simplicus in F. M. Cornford, "Innumerable Worlds in Pre-Socratic Philosophy", *Classical Quarterly* **28**, 13 (1934). Also see Michael J. Crowe, *The Extraterrestrial Life Debate 1750–1900*, Cambridge University Press, New York 1986, page 4.

25. Lucretius (99–55 BC), *De rerum natura* (*The Nature of the Universe*), 55 B.C. English translation by Ronald E. Latham, Penguin Books, New York, 1951, page 91.

26. Giordano Bruno (1548–1600), *De l'infinito universo et mondi* (*On the Infinite Universe and Worlds*) 1584. Third dialogue between Burchio, Elpino, Fracastoro and Philotheo. English translation by Dorothea Waley Singer, published by Henry Schuman, New York 1950.

27. Michel Mayor and Pierre-Yves Frei, *New Worlds in the Cosmos: The Discovery of Exoplanets*, Cambridge University Press 2003, pages 18.

28. Bob Dylan (1941–), *The Times They Are A'Changin*.

29. John Milton (1608–1674), *Il Penseroso* 1, 65.

30. Samuel Taylor Coleridge (1772–1834), *The Rime of the Ancient Mariner*.

31. Simplicus' *Commentary*, quoted in Pierre Duhem's *To Save The Phenomena: An Essay on the Idea of Physical Theory from Plato to Galileo*. English translation by Edmund Doland and Chaninah Maschler. University of Chicago Press, Chicago, Illinois 1969, page 5.

32. Henry Wadsworth Longfellow (1807–1882), *Evangeline, Part One: III*.

33. John Maynard Keynes (1883–1946), "Newton, The Man". In *The Royal Society Newton Tercentenary Celebrations 15 July 1946*, Cambridge, Printed for the Royal Society (London) by the University Press 1947, pages 27–34. Keynes gave his talk at Cambridge in 1942 and died in 1946, leaving the manuscript, which was read by his brother Geoffrey Keynes at the Royal Society celebrations.

34. Marcus Aurelius (161–180), *Meditations*.

35. Francis William Bourdillon (1852–1921), *The Night Has a Thousand Eyes*.

36. John Milton (1608–1674), *Paradise Lost, Book VII, Line 577*.

37. Camille Flammarion (1842–1925), *Les Etoiles*, Paris, 1882, page 181. English quotation by Charles A. Whitney in "The Skies of Vincent Van Gogh," *Art History* **9**, 358 (1986).

38. George Ellery Hale (1868–1938), *Harper's Magazine* **156**, 639–646 (1928).

39. Eihei Dogen (1200–1253), *Death Poem*.

40. Hans A. Bethe (1906–2005), "My Life in Astrophysics", *Annual Review of Astronomy and Astrophysics* **41**, 6-7 (2003). Mr. Rydberg of the Nobel Foundation, who met Bethe when he arrived to receive the 1967 Nobel Prize in Physics, told Bethe the reason for excluding astronomers and pure mathematicians in the Nobel Prize bequest.

41. Edwin P. Hubble (1889–1953), *The Realm of the Nebulae*, Yale University Press, New Haven 1936, pages 201, 202.

42. William Butler Yeats (1865–1939), *The Second Coming*.

43. Robinson Jeffers (1887–1962), *The Great Explosion*.

44. Martin Ryle (1918–1984). As remembered by Thomas Gold (1920–2004) and quoted by Simon Singh in *Big Bang – The Origin of the Universe*, New York, Forth Estate/ Harper Collins 2004, page 414.

45. Walter Baade (1893–1960) and Fritz Zwicky (1898–1974), "Cosmic Rays from Super-Novae," *Proceedings of the National Academy of Sciences* **20**, No. 5, 259–263 (1934).

46. John Michell (1724–1793). "On the means of discovering the distance, magnitude, etc., of the fixed stars, in consequence of the diminution of their light, in case such

a diminution should be found to take place in any of them, and such other data should be procured from observations, as would be further necessary for that purpose." *Philosophical Transactions of the Royal Society (London)* **74,** 35 (1784).

47. Robinson Jeffers (1887–1962), *The Great Explosion.*

48. Georges Lemaître (1894–1966), in an article entitled "L'expansion de l'espace" published in *La Revue des Questions Scientifiques, 4e Série.* November 1931. English translation in *The Primeval Atom: An Essay on Cosmogony* by Georges Lemaître, D. Van Nostrand Co., New York 1950, page 78.

49. Robert Wilson (1933–), interview in 1982, reported by Jeremy Bernstein in *Three Degrees Above Zero,* New York, Scribner's 1984, page 205. Also see his *New Yorker* article of the same title on 20 August 1984.

50. Claude Debussy (1862–1918), *Proses lyriques, Of the dream.*

51. *The Music of the Night,* lyrics by Charles Hart and music by Andrew Lloyd Webber, in *The Phantom of the Opera,* 1986.

52. Henry Wadsworth Longfellow (1807–1882), *The Day is Done.*

53. Paul Klee (1879–1940). In 1908 Klee wrote: "I cannot find sleep. In me the fire still glows, in me it still burns here and there. Seeking a breath of fresh air, I go to the window and see all the lights darkened outside. Only very far away a small window is still lit. Is not another like me sitting there? There must be some place where I am not completely alone! And now the strains of an old piano reach me, the moans of the other wounded person." In: Felix Klee (editor), *The Diaries of Paul Klee 1898–1918,* Berkeley, University of California Press 1964, Letter 833, page 229.

54. Bob Dylan (1941–), *She Belongs to Me.*

55. Hermann Minkowski (1864–1909), *Space and Time.* An address delivered at the 80th Assembly of German Natural Scientists and Physicians at Cologne, 21 September 1909. English translation in *The Principle of Relativity* (editor Arnold Sommerfeld). London, Methuen 1923, reprinted Dover, New York 1942, page 75.

56. Alan Lightman (1948–), *Einstein's Dreams,* New York, Random House, Warner Books 1993, pages 70 to 72.

57. Einstein's conversations or letters quoted by Abraham Pais in '*Subtle is the Lord': The Science and the Life of Albert Einstein,* New York, Oxford University Press 1982, page 253.

58. Albert Einstein (1879–1955), "Explanation of the Perihelion Motion of Mercury by Means of the General Theory of Relativity," *Sitzungsberichte der Preussischen Akademie der Wissenschaften zu Berlin* **11,** 831–839 (1915). English translation in Kenneth R. Lang and Owen Gingerich (editors), *Source Book in Astronomy and Astrophysics 1900–1975,* Cambridge, Massachusetts, Harvard University Press 1979, page 822.

59. A. Vibrant Douglas, *The Life of Arthur Stanley Eddington,* London, Thomas Nelson and Sons 1956, page 40.

60. Quoted in R. Clark, *Einstein: The Life and Times,* New York, Avon Books 1984, page 287.

61. T. S. Eliot (1888–1965), *Four Quartets,* "*East Coker,*" III, I.

62. William Shakespeare (1564–1616), *Henry V, Act IV, Scene 1.*

63. Agnes M. Clerke (1842–1907), *Problems in Astrophysics.* Adam and Charles Black, London 1903, page 400.

64. Fritz Zwicky (1898–1974), "Die Rotverschiebung von extragalaktischen Nebeln," *Helvetica Physica Acta* **6,** 110 (1933). English translation by Sydney van den Berg in "The Early History of Dark Matter", *Publications of the Astronomical Society of the Pacific* **111,** 657 (1999).

65. Georges Lemaître (1894–1966), "The beginning of the world from the point of view of quantum theory," *Nature* **127,** No. 3210, 706 (1931).

66. Alan H. Guth (1947–), *The Inflationary Universe: The Quest For A New Theory of Cosmic Origins,* New York, Addison-Wesley 1997, page 15.

67. John Maddox (1925–), *What Remains to be Discovered,* Simon and Schuster, New York 1998, page 47.

68. Werner Heisenberg (1901–1976), *Physics and Philosophy,* New York, Harper and Row, 1958, 1962, page 58.

69. Sheldon Glashow (1932–), Interview on *The Elegant Universe,* a NOVA television program presented by the Public Broadcasting System in 2004.

70. Woody Allen (1935–), "Strung Out", *The New Yorker*, 28 July 2003, page 96.

71. Brian Greene (1963–), *The Elegant Universe: Superstrings, Hidden Dimensions and the Quest for the Ultimate Theory*, New York, W. W. Norton 1999, pages 18, 19.

72. Arthur Eddington (1882–1944), *The Expanding Universe*, Cambridge University Press, Cambridge, England 1933, pages 24 and 56.

73. Georges Lemaître (1894–1966), in an article entitled "L'expansion de l'espace", published in *La Revue des Questions Scientifiques*, 4e Série, November 1931. English translation in *The Primeval Atom: An Essay on Cosmogony* by Georges Lemaître, D. Van Nostrand Co., New York 1950, page 79.

74. Buddha (560–483 BC), In *Chapters Spoke with Intention (Udanavarga)*.

75. Thomas Carlyle (1795–1881), *On Heroes, Hero Worship and the Heroic in History*, Lecture 4.

76. Julian Huxley (1887–1975), *Cosmic Death*.

77. Robinson Jeffers (1887–1962), *Margrave*.

78. Albert Einstein (1879–1955). Letter to Max Born, December 4, 1926. In Irene Born (translator), *The Born-Einstein Letters*, New York, Walker 1971, page 91. Reproduced in Alice Calaprice (editor), *The Expanded Quotable Einstein*, Princeton, New Jersey, Princeton University Press 2000, page 245.

79. Mark Twain (1835–1910), *Life on the Mississippi*.

80. Plato (c. 420–340 BC), *The Republic* VII, 52.

81. Paul Davies (1946–), "Wormholes through physics" *Nature* **413**, 354–355 (2001) September 27, page 354.

82. Stephen W. Hawking (1942–), *A Brief History of Time: From the Big Bang to Black Holes*, New York, Bantam Books 1988, page 195.

83. Saint Augustine (354–430), *De Genesi ad Literam*, or *The Literal Meaning of Genesis*. English translation by John Hammond Taylor, Volume 41 in the series, *Ancient Christian Writers*, New York, Newman Press 1982, Book four, chapter 35 page 145 and book five, chapter 5 page 153.

84. Howard J. Van Till (1938–), "Basil, Augustine, and the Doctrine of Creation's Functional Integrity," *Science and Christian Belief*, Vol. 8, No. 1, page 32 (April, 1996). One of the chief works considered in that essay is: St. Augustine (354–430), *De Genesi ad Litteram*, or *The Literal Meaning of Genesis*. English translation by John Hammond Taylor, volumes 41 and 42 in the series, *Ancient Christian Writers*, New York, Newman Press 1982.

85. Pope Pius XII (1876–1958), *Un' Ora di serena letizia*. Address in Italian delivered to the Pontifical Academy of Science on 22 November 1952. English translation by P. J. McLaughlin in *The Church and Modern Science*, New York, Philosophical Library 1957. The address includes: "If we direct our attention to the past, the farther back we go, the more matter presents itself as rich in free energy and a theatre of great cosmic events. Everything seems to indicate that the material content of the Universe had a mighty beginning in time, being endowed at birth with vast reserves of energy, in virtue of which, at first rapidly, and then ever more slowly, it evolved into its present state… Thus, with that concreteness which is characteristic of physical proofs, it [modern science] has confirmed the contingency of the Universe and also the well-founded deduction as to the epoch when the world came forth from the hands of the Creator. Hence, creation took place. We say: therefore, there is a Creator. Therefore, God exists!"

86. Albert Einstein (1879–1955), *Cosmic Religion*, New York, Covici Friede 1931.

87. Alexander Pope (1688–1744), *An Essay on Man* (1733), Epistle I, lines 23 to 28.

88. T. S. Eliot (1888–1965), *The Hollow Men*.

89. Robert Frost (1874–1963), *Fire and Ice*.

90. T. S. Eliot (1888–1965), *Four Quartets*, "*Little Gidding*", V, lines 26 to 29.

91. Mario Socrate (1920–), *Favole parabliche* or *Parabolic Fables*.

FURTHER READING

Armitage, Angus: *Edmond Halley*. Thomas Nelson and Sons, London 1966.

Bartusiak, Marcia: *Einstein's Unfinished Symphony: Listening to the Sounds of Space-Time*, Berkley Books, New York 2000.

Bartusiak, Marcia: *Through a Universe Darkly*. Avon Books, New York 1993.

Bryson, Bill: *A Short History of Nearly Everything*. Broadway Books, New York 2003.

Cook, Alan: *Edmond Halley: Charting the Heavens and the Seas*. Clarendon Press, Oxford 1998.

Croswell, Ken: *The Universe at Midnight: Observations Illuminating the Cosmos*. Simon and Shuster, New York 2001.

Croswell, Ken: *The Alchemy of the Universe: Searching for Meaning in the Milky Way*. Anchor Books, Doubleday, New York 1995.

Crowe, Michael J.: *The Extraterrestrial Life Debate 1750–1900: The Idea of a Plurality of Worlds From Kant to Lowell*. Cambridge University Press, New York 1986.

DeVorkin, David H.: *Science With A Vengeance: How the Military Created the US Space Sciences After World War II*. Springer Verlag, New York 1992.

Drake, Stillman: *Discoveries and Opinions of Galileo. Translated with an Introduction and Notes*, Doubleday Anchor, New York 1957.

Duhem, Pierre: *To Save The Phenomena: An Essay on the Idea of Physical Theory from Plato to Galileo*. The University of Chicago Press, Chicago 1969.

Dyson, Freeman J.: *The Sun, the Genome, the Internet*. Oxford University Press, New York 1999.

Eddington, Arthur: *The Expanding Universe*. Cambridge University Press, Cambridge, England 1933.

Ferrier, Jean-Louis: *Paul Klee*. Pierre Terrail Editions, Paris 1999.

Ferris, Timothy: *Coming of Age in the Milky Way*. William Morrow and Company: New York 1988.

Ferris, Timothy: *The Whole Shebang. A State-of-the Universe(s) Report*. Simon & Shuster, New York 1997.

Galilei, Galileo: *Sidereus Nuncius or The Sidereal Messenger,* Translation with Introduction, Conclusion, and Notes by Albert Van Helden, The University of Chicago Press, Chicago 1989. Also see *Discoveries and Opinons of Galileo. Translated with an Introduction and Notes by Stillman Drake*. Doubleday Anchor, New York 1957.

Greene, Brian: *The Elegant Universe: Superstrings, Hidden Dimensions, and the Quest for the Ultimate Theory*. W.W. Norton, New York 1999.

Guth, Alan H.: *The Inflationary Universe: The Quest for a New Theory of Cosmic Origin:* Helix Books, Addison-Wesley, New York 1997.

Hahl-Koch, Jelena: *Kandinsky,* Rizzoli, New York 1993.

Hirshfeld, Alan W.: *Parallax: The Race to Measure the Cosmos,* W.H. Freeman and Co., New York 2001.

Hoskin, Michael A. (ed.): *The Cambridge Illustrated History of Astronomy,* Cambridge University Press, New York 1997.

Kandinsky, Wassily: *Concerning the Spiritual in Art.* Original German edition *über das Geistige in der Kunst,* published in1912. English translation published by George Wittenborn, Inc., New York, 1947.

Kirshner, Robert P.: *The Extravagant Universe,* Princeton University Press, Princeton, New Jersey 2002.

Klee, Felix (ed.): *The Diaries of Paul Klee: 1898–1918.* Berkeley, University of California Press 1964.

Kragh, Helge: *Cosmology and Controversy: The Historical Development of Two Theories of the Universe.* Princeton University Press, Princeton, New Jersey 1996.

Lang, Kenneth R.: *Companion to Astronomy and Astrophysics: Chronology and Glossary with Data Tables.* Springer Verlag, New York 2006.

Lang, Kenneth R.: *The Cambridge Encyclopedia of the Sun,* Cambridge University Press, New York 2001.

Lang, Kenneth R.: *The Sun From Space,* Springer Verlag, New York 2000.

Lang, Kenneth R.: *Astrophysical Formulae. Volume II. Space, Time, Matter and Cosmology. Third Enlarged and Revised Edition.* Springer Verlag, New York 1999.

Lang, Kenneth R, and Gingerich, Owen (eds): *A Source Book in Astronomy and Astrophysics, 1900–1975.* Harvard University Press, Cambridge, Massachusetts, 1979.

Lemaître, Canon Georges: *The Primeval Atoms. An Essay on Cosmogony.* D. Van Nostrand Co, New York 1950.

Lindsay, Kenneth C. and Vergo, Peter (eds.): *Kandinsky: Complete Writings on Art.* G. K. Hall & Co.: Boston, Mass. 1982, Reprinted in paperback by Da Capo Press, New York 1994.

Maddox, John: *What Remains to be Discovered.* The Free Press, New York 1998.

Matisse, Henri: *Jazz.* First published in 1947 by Tériade, Paris. English translation published by Prestel Verlag, New York 2001.

Mayor, Michel and Frei, Pierre-Yves, *New Worlds in the Cosmos: The Discovery of Exoplanets*. Cambridge University Press, New York 2003.

Panek, Richard: *Seeing and Believing: How the Telescope Opened Our Eyes and Minds to the Heavens*, Viking Penguin, New York 1998.

Porter, Roy and Ogilvie, Marilyn (eds.): *The Biographical Dictionary of Scientists. Third Edition.* Oxford University Press, New York 2000.

Rees, Martin: *Before the Beginning: Our Universe and Others*. Addison-Wesley, Reading, Massachusetts 1997.

Ricard, Matthieu and Trinh Xuan Thuan: *The Quantum and the Lotus: A Journey to the Frontiers where Science and Buddhism Meet*, translation by Ian Monk, Crown Publishers, New York 2001.

Schilling, Govert: *Flash! The Hunt for the Biggest Explosions in the Universe*. Cambridge University Press, New York 2002.

Singh, Simon: *Big Bang: The Origin of the Universe*, Fourth Estate/Harper Collins, New York 2004.

Sobel, Dava: *Longitude: The True Story of a Lone Genius Who Solved the Greatest Scientific Problem of His Time*, Walker and Company, New York 1995.

Thorne, Kip S.: *Black Holes and Time Warps: Einstein's Outrageous Legacy*. W.W. Norton & Co., New York 1994.

Van Gogh, Vincent: *The Complete Letters of Vincent Van Gogh*. New York Graphic Society, Greenwich, Connecticut 1958.

Van Helden, Albert: *Galilei, Galileo: Sidereus Nuncius or The Sidereal Messenger, Translation with Introduction, Conclusion and Notes*. The University of Chicago Press, Chicago 1989.

Wilford, John Noble (ed.): *Cosmic Dispatches. The New York Times Reports on Astronomy and Cosmology*. W.W. Norton, New York 2001.

Wright, Thomas: *An Original Theory or New Hypothesis of the Universe*. First published in 1750. Reproduced by Michael A. Hoskin (ed.) Neale Watson Academic Publications, New York 1971.

Author Index

Subject Index

CPSIA information can be obtained
at www.ICGtesting.com
Printed in the USA
LVHW080942021218
598912LV00017B/8/P